Contents

WBI LEARNING RESOURCES SERIES

Knowledge Economies in the Middle East and North Africa

Toward New Development Strategies

Edited by
Jean-Eric Aubert
Jean-Louis Reiffers

The World Bank
Washington, D.C.

Library of Congress Cataloging-in-Publication Data

World Bank Forum on Knowledge for Development in the Middle East and North Africa
   (2002 : Marseille, France)
   Knowledge economies in the Middle East and North Africa : toward new development
strategies /edited by Jean-Eric Aubert, Jean-louis Reiffers.
      p.   cm. — (WBI development studies)
   Papers from the World Bank Forum on Knowledge for Development in the Middle East
and North Africa, Sept. 9–12, 2002, Marseilles, France.
   Includes bibliographical references.
   ISBN 0-8213-5701-8
      1. Knowledge management—Middle East—Congresses. 2. Knowledge
management—Africa, North—Congresses. 3. Information technology—Middle
East—Management—Congresses. 4. Information technology—Africa,
North—Management—Congresses. 5. Technological innovations—Economic
aspects—Middle East—Congresses. 6. Technological innovations—Economic
aspects—Africa, North—Congresses. 7. Technology and state—Middle East—Congresses. 8.
Technology and state—Africa, North—Congresses. 9. Education and state—Middle
East—Congresses. 10. Education and state—Africa, North—Congresses. 11. Middle
East—Economic conditions—1979—Congresses. 12. Africa, North—Economic
conditions—Congresses. I. Aubert, Jean-Eric. II. Reiffers, Jean-Louis III. Title. IV. Series.

HD30.2.W67 2002
303.48'33'0956—dc22

                                                                      2003064599

Boxes

Figures

Tables

Foreword

The Middle East and North Africa (MENA) region has been facing considerable economic challenges. Left behind by the industrial revolution, overly dependent on oil resources, and on the fringes of the globalization process, a number of MENA countries have embarked on structural reforms to overcome economic stagnation, mounting unemployment, and increasing poverty.

At the same time, there is growing awareness worldwide that the knowledge revolution offers new opportunities for growth resulting from the availability of information and communication technologies and from the advent of a new form of global economic development rooted in the concept of the knowledge economy, which is based on the creation, acquisition, distribution, and use of knowledge.

Senior government officials, distinguished leaders from the private sector, and members of civil society from the Middle East and North Africa region met in Marseilles, France, on September 9–12, 2002, to discuss policy strategies for tapping into the growth opportunities offered by the knowledge economy. Policymakers from 13 countries and some 50 scholars from around the world examined national experiences and exemplary initiatives in such key sectors as education, science and technology, and information infrastructure that may hold promise for the region. The conference, which was organized by the World Bank, offered an opportunity for dialogue and exchange not only on policy issues, but also on the practical know-how that has emerged from successful solutions in the region and in other parts of the world.

This book results from the papers prepared for discussion during the conference. It provides quantitative analysis to help benchmark the countries against worldwide knowledge economy trends, identifies key implementation issues, and presents relevant policy experiences.

On behalf of the World Bank Institute and the World Bank's Middle East and North Africa Region I would like to thank the city of Marseilles for hosting the conference and for its standing offer to host others in the coming years. Activities such as this one will help us share lessons learned in preparing for the knowledge economy, monitor trends worldwide and analyze their effects on the developing world, and continue to disseminate good policy practices.

Frannie A. Léautier
Vice President
World Bank Institute

Acknowledgments

The conference "Knowledge for Development: A Forum for Middle East and North Africa" was held in Marseilles, France, September 9–12, 2002. It was organized by the World Bank's Middle East and North Africa region and the World Bank Institute in cooperation with the City of Marseilles. This volume was prepared by Jean-Eric Aubert (World Bank Institute) and Jean-Louis Reiffers (Institut de la Méditerranée) based on the background report prepared for the conference and the contributions of conference participants.

The authors benefited from the support of Frédéric Blanc and Clotilde Boutrolle (Institut de la Méditerranée) and Marie-Hélène Fandel (World Bank). Specific information was provided by Anna Bjerde (Middle East and North Africa [MENA] region) for chapter 3; André Kirchberger (consultant) for chapter 6; Abdelkader Djeflat (consultant) for chapter 7; and Samia Melhem (Global Information and Communication Technologies Department, World Bank) for chapter 8. Carl Dahlman and Chantal Dejou (World Bank Institute); Emmanuel Forestier, Nadereh Chamlou, Omer Karasapan, Françoise Clottes, and Regina Bendokat (MENA); and Jean-Claude Tourret (Institut de la Méditerranée) made valuable comments.

On January 13, 2003, the report was reviewed at a meeting chaired by Mustapha Kamel Nabli, Chief Economist, MENA Region. The reviewers were Farrukh Iqbal (MENA) and Mostafa Terrab (Global Information and Communication Technologies Department, World Bank), whose helpful comments are gratefully acknowledged.

Preface

For the greater part of the last millennium, the countries of the Middle East and North Africa (MENA) region were socially, economically, and technologically very advanced. Beyond their differences, they shared an openness to knowledge and to other cultures. They represented the predominant civilization from the 8th to the 13th century and remained powerful until the 18th century.

These countries made innumerable contributions in mathematics, astronomy, medicine, architecture, and philosophy,[1] and the Islamic world was the main "global" provider of knowledge. It reactivated and added to the discoveries of Greece and Rome, India, and China. The hunger for knowledge does not, by itself, account for the golden age of Islam, but it was a major driving force. Progress on the intellectual front was paralleled by the development of an economy based on a monetary market and commerce, along with technological advances such as better and faster ships to service the merchant fleets.

The region declined because it was unable to find its place in an economy of iron, coal, and steam, an inability largely shared by all Mediterranean countries. Scientific thought and the principles of the modern scientific method—based on the questioning of dogma and on systematic experimental research—shifted to northern Europe. This is likely one reason why the MENA region, and the Mediterranean region more generally, dropped behind. It suffered the same fate as other regions bypassed by the Industrial Revolution and suffered a serious setback lasting over two centuries.

At their independence, most MENA countries became dependent on oil and other natural resources. Encouraged by galloping demographics, they invested income from the oil boom in large infrastructure projects, education, and public health. Their economic and institutional

---

1. The Baghdad "House of Wisdom" founded by Al-Ma'mun, the second son of the Abassid Caliph Harun al-Rashid (813–833), offers a striking example of this flourishing civilization. This is where Greek manuscripts were collected and translated and where scholars were welcomed, where the scientist and philosopher Al-Kindi discovered links between Platonic philosophy and Islamic thought, where Al-Khawarizmi conducted research on astronomy and mathematics (astronomical tables, an improvement on the Greek astrolabe), where Al-Battani worked on his book of astronomical tables (his contributions were such that the giants of early modern astronomy—Copernicus, Kepler, Galileo—referred to them in their work). In the field of pure mathematics, Arab contributions to algebra and trigonometry were fundamental: Al-Battani founded trigonometry (the sine and cosine functions), and algebra was established as an autonomous discipline by Al-Khawarizmi and geometric algebra by Al-Tûsï. Many stories have been told about the works of Al-Kindi, the first great philosopher of Islam, or about the genius of Avicenna, the medical doctor and thinker who, together with Al-Farabi (of Turkestan) and Averroes (of Cordoba) created the chain of Muslims who took Greek philosophy and ancient wisdom as their starting point and made their contributions to modern philosophical thought.

frameworks, however, left much to be desired in terms of governance, competition, and transparency. Economic performance gradually declined, and poverty increased significantly throughout the region.

Today a slow change is under way. These countries are gradually strengthening their economic systems, but they are not keeping up with the pace in the rest of the world and cannot hope to do so under present conditions. To progress, the MENA region must face its internal challenges and become part of the knowledge and information revolution, which is likely to lead to very rapid changes worldwide. Countries that fail to become part of this revolution risk becoming even more marginalized than those left aside in the earlier industrial revolution.[2]

If the region's governments wish to take advantage of opportunities that lie within their reach, they must become the architects of new institutions and the promoters and regulators of economies based on the purchase, production, and dissemination of knowledge.

Following a brief overview (chapter 1), the discussion begins by making more explicit the challenges confronting the region's countries (chapter 2), before describing the main characteristics of the knowledge-based economy that is taking shape worldwide and offers MENA countries a way forward if they take judicious measures to meet their needs and benefit from their capabilities (chapter 3). Chapter 4 then analyzes MENA countries' readiness for the knowledge economy on the basis of a set of indicators. Chapter 5 discusses prerequisites and the fundamentals of the economic and institutional framework. The basic policy elements of knowledge-based economy strategies are detailed: the renovation of education systems (chapter 6), the creation of a climate conducive to innovation (chapter 7), and the development of an efficient telecommunications infrastructure as the foundation of a new era (chapter 8). In chapter 9, the formulation of national visions and strategies is discussed. A brief conclusion follows (chapter 10). Examples from the MENA region and other parts of the world illustrate the various chapters. A set of data that makes it possible to benchmark and position MENA countries' readiness for the knowledge economy is presented in an appendix.

The purpose of this book, which is largely based on information gathered at the conference in Marseilles, is to set out relevant issues and provide general policy orientations.[3] Thus, the analysis remains at a certain level of generality. It constitutes a basis for undertaking more documented, in-depth and empirical work.

---

2. Forceful and carefully documented analyses of the problems encountered by the MENA region can be found in UNDP (2002) and World Economic Forum (2003).

3. A CD-ROM containing all the documents presented at the conference as well as related information (such as relevant websites) is available. The documents can also be consulted at the conference website (www.worldbank.org/k4dmarseille).

1

Overview

This book analyzes the development of knowledge-based economies in the Middle East and North Africa (MENA). Its principal messages are:

- Because of the so-called "knowledge revolution" resulting from the rapid growth in information and communication technologies (ICT), the acceleration of technical change and the intensification of globalization, a new form of economic development is taking shape worldwide.
- The knowledge revolution presents MENA countries with challenges and opportunities. They need to take advantage of this new source of growth and employment. To date, related investments in education, information infrastructure, research and development (R&D), and innovation have been insufficient or inappropriate in most MENA countries. Moreover, inadequate economic and institutional frameworks prevent these investments from yielding desired results.
- MENA countries risk falling further behind in the world economy. Urgent action is needed to advance structural reform and to intensify and adapt knowledge-related investments.

These messages concur with those of two important recent reports on Arab economies by the United Nations Development Programme (UNDP, 2002) and the World Economic Forum (2003). While there seems to be agreement on what needs to be done in the region, the question of how to achieve the desired results is unfortunately often left unexplored. This is to be the focus of further World Bank conferences.

The Need for a New Form of Development

The MENA region was a source of global innovation and modernization at the beginning of the last millennium. Left behind by the Industrial Revolution, the region's socioeconomic situation has gradually worsened. Today, economic stagnation and mounting unemployment are of concern in the region and in the international community.

The MENA countries need a more productive economic regime to boost development and provide the 40 million jobs that must be created over the next 10 years to absorb increases in the labor force. With unemployment currently at some 20 percent of the labor force in the region as a whole (with the exception of the United Arab Emirates), there is a serious risk of social and political instability.

The participation of MENA countries in the so-called knowledge revolution presents challenges but also offers a unique opportunity to evolve in a direction better suited to their current and future socioeconomic needs.

A Knowledge-Based Development Process

Knowledge has always been the source of economic development. Economies that perform effectively have been and continue to be those that make the best use of knowledge and its applications. Important changes over the past decade have however given rise to the term "knowledge revolution."

The knowledge revolution results principally from an intensification of the globalization process, the spread of ICT, more generalized automation and computerization of productive activities, the increasingly tight links between science and innovation, and the development of new fields such as biotechnologies. These changes are dramatically altering business climates and the conditions of economic growth and competitiveness worldwide. The economy that is taking shape is captured by the expression "knowledge-based economy" or "knowledge economy." In such an economy, knowledge enriches all sectors and agents. It is a source of new industries and of renewal of established ones and a key factor in competitiveness and social welfare.

The World Bank has developed a four-dimensional framework that captures the fundamental elements of the knowledge economy and makes it possible to gauge and compare countries' progress in becoming part of the knowledge economy. Its "pillars" are an economic and institutional regime that encourages efficient use of knowledge, the flourishing of entrepreneurship, an educated, creative, and skilled population, a well-developed information and communication infrastructure, and an effective innovation system with dynamic interaction between the world of science and technology and the world of business. In addition, a fifth pillar is constituted by the intangible ingredients of a cultural nature that relate to collective trust and vision and determine a society's inner dynamism.

Readiness for the Knowledge Economy

The MENA region's readiness for the knowledge economy is low, although a number of governments have begun to adapt their economies to meet the new challenges. Compared to other parts of the developing world, the region trails East Asia, Eastern Europe, Central Asia, and Latin America. It is somewhat ahead of South Asia and Sub-Saharan Africa. In general terms, the MENA countries' knowledge economy readiness is somewhat lower than their overall level of economic development as measured by gross domestic product (GDP).

There are, however, important differences within the region. Jordan and the United Arab Emirates have shown the way forward and have made important reforms and investments, while other countries have been more timid or less systematic.

The Economic and Institutional Regime

An appropriate economic and institutional regime is essential for ensuring proper payoff from investments in knowledge, information, education, and research. Key elements of such a regime are the competitive environment, financial markets, labor markets, safety nets, legal systems, and, more generally, the overall governance climate.

The MENA region suffers from an over-large public sector, over-regulation, bureaucracy, and control of information. Along with greater freedom and civil rights, individuals and organizations would benefit from more autonomy so that entrepreneurship and innovative behavior can flourish at all levels and in all sectors. Some countries in the region have already used e-government to promote transparency and availability of information.

An inappropriate regulatory and legal framework is the principal cause of the low level of foreign direct investment (FDI). At 6 percent of GDP, the level of FDI is the world's lowest. This seriously hinders inflows of knowledge and technology. Economic openness has also suffered from high tariff and nontariff barriers, but the situation is improving. Association agreements have been signed with advanced economies as well as for trade integration within the region.

Poorly functioning financial and capital markets affect the region's innovation and modernization capabilities. Banks do not operate freely (they are obliged to support public enterprises), stock exchanges are underdeveloped, foreign exchange is controlled. This is an area in which structural reforms need to be actively pursued, while taking advantage of opportunities offered by the specific characteristics of Islamic banking and related risk-sharing practices.

Human Resources

Considerable efforts made in education, measured as a share of GDP (over 5 percent), have contributed significantly to reducing illiteracy. However, the quality of education needs to be improved at all levels to train the labor force to meet the region's economic needs. At present, the job market does not provide the educated population with appropriate employment opportunities, and this has led to high unemployment among diploma holders and to a significant brain drain.

Education is still largely based on rote learning, which relies on memorization and does not stimulate creativity. Extensive "arabization" of education would help raise the population's basic literacy levels and strengthen its identity, but early learning of other major international languages is also essential for global trade and for scientific and other exchanges.

Systems of lifelong learning also need to be put in place. This requires developing curricula that address the qualifications and skills needed by the work force as these evolve over time.

Finally, there is the problematic status of women in a number of these countries. In neglecting basic education for women (50 percent are illiterate), in excluding them from the labor force (less than 25 percent of the labor force in the region as a whole), and in barring them from political life, these societies deprive themselves of a considerable source of dynamism, creativity, and knowledge.

Innovation

Efforts have been made to innovate, with some noteworthy successes. Yet the modernization of traditional sectors and support for emerging industries remain insufficient. R&D represents only 0.3 percent of GDP, and technological and managerial modernization suffers from the small volume of FDI.

Effective interaction between institutions of science and education and the business world is essential to a dynamic innovation climate, as is a flexible framework in which entrepreneurial initiatives can flourish. This requires a supportive environment for innovators in search of financial, commercial, and technical support, as well as quality certification and standards that can lead to the growth of more efficient businesses.

MENA countries have good research institutes in certain fields. Too often, however, R&D institutions are constrained by rigid regulations, lack of budget autonomy, and lack of incentives to collaborate with industry. Governments should undertake the necessary reforms and increase expenditures on R&D. The private sector must also contribute, but this requires a more favorable climate for innovation. Strategies need to be defined for building on the region's comparative advantages by developing oil-related technologies or technologies that address the region's needs such as water.

However, domestic R&D is not the principal source of knowledge for developing countries. In addition to FDI, it is crucial to tap into global knowledge and technology through appropriate trade, licensing, and technology-import policies as well as through efficient links to highly qualified expatriates from MENA countries in advanced economies (of which there are more than a million).

Telecommunications and Information Infrastructure

The region, as a whole, has made progress in the access to and use of ICT. However, teledensity levels (number of telephones per inhabitant) remain below 10 percent in most countries. The Internet is little used throughout the region (less than 1 percent of the world's total user base).

A number of MENA countries have recently adopted telecommunications measures that follow global good policy practices—establishment of an independent authority, breaking up of public monopolies, liberalization, and so on. However, the reforms are often incomplete. Moreover, governments may need to invest more, notably when telephone companies, affected by the economic slowdown, are reluctant to invest.

Rapid growth in mobile phones has compensated for the relative underequipment in fixed lines but has, at the same time, slowed the establishment of the infrastructure needed for the Internet, which is used by less than 1 percent of the population in most MENA countries. In addition, MENA countries need to develop Arabic content for ICT applications and to ensure that the general population has the skills necessary to use ICT effectively.

New Visions and Strategies

To move toward knowledge-based development strategies, some countries in the region have implemented initiatives and reforms that are gradually bearing fruit. They provide examples to emulate, and inspiring models can be found in other parts of the world as well. Country leaders and communities concerned with change and progress need to take action and promote efficient policies.

A country's rulers and leaders can articulate and propose a vision of a new form of development, with freedom in a central place and a society well integrated in the world economy. Strategies need to address the four basic pillars, and concrete, visible achievements help build trust. The push toward knowledge-based development could be facilitated by establishing powerful interministerial agencies able to set priorities and engage in budget arbitration.

More than further government investment (increasingly difficult in any event given mounting budget constraints), the region needs new approaches to governing. Much can be gained through well-targeted institutional and regulatory action that can mobilize or release human energies and resources that are currently trapped.

Cooperation among MENA countries is crucial for activating change and progress. Given the region's extensive linguistic homogeneity, trade and commercial exchanges are limited but can certainly be increased. Similarly, cooperation can be developed in science, education, and culture, using new technological opportunities and approaches such as open and virtual universities. Joint investment in (broadband) telecommunications would also facilitate integration. The sense of common identity within the Arab-Islamic world would facilitate such initiatives. International organizations, including the World Bank, can play a crucial role in accompanying these developments. New forms of support can help local capacity-enhancing processes.

2

The Challenge: Changing the Growth Model

A Distorted Development Process

Since World War II, economic growth in the MENA region has been mainly driven by the exploitation and exportation of natural resources, mostly oil. Benefiting from the boom in oil prices, MENA countries enjoyed high growth rates in the 1970s and 1980s and invested heavily in ambitious development projects, education, and public health.

This allowed them to reach a reasonable level of development (see table 2.1), as measured by UNDP's human development indicators, which add to common economic indicators, such as GDP per capita, information on literacy and health. The MENA region is still in a better position than Sub-Saharan Africa, South Asia, and East Asia (including China).

However, growth rates fell significantly in the 1990s. In fact, over the period 1965–2000, the region had the second lowest growth rate worldwide (3 percent). Productivity fell at 5.66 percent, the fastest rate of all World Bank Regions (table 2.1). Unemployment stands at more than 20 percent of the labor force. Various features indicate that the development process has reached certain limits (Dahlman, 2002):

- Dependence on oil, tourism and remittances and very limited development of the manufacturing sector.
- Little diversification of export products. As a whole, MENA has the lowest share of manufactured exports in nonoil sectors (14 percent) and the highest share of fuel exports (80 percent). To give an order of magnitude, the region's total nonoil exports (for a population of 300 million) are lower than Finland's (for a population of 5 million).
- Poor integration in the world economy. The region's average rate of FDI is 6 percent of GDP, the world's lowest.
- An insufficiently developed private sector. The public sector remains the main source of jobs (public administrations, state enterprises).

Table 2.1. *Selected Development Indicators by Main World Bank Regions*

Region	Average annual growth in GDP 1990–99 (%) (WDI, 2002)	Productivity growth (percentage change in GDP per person employed) 2000 (IMD, 2001)	Human development index 1999 (UNDP, 2002)	Gender development index 1999 (UNDP, 2002)	Poverty index 1999 (UNDP, 2002)
MENA	4.01	−5.66	0.71	0.71	24.54
South Asia	4.84	4.10	0.56	0.54	35.50
East Asia	5.29	1.88	0.77	0.77	15.78
Africa	3.42	6.78	0.48	0.48	34.75
Latin America Caribbean	3.10	−0.24	0.74	0.72	14.58

In addition, the region has the least water and second smallest amount of arable land of all world regions. At the same time, having been more or less continuously involved in conflicts for more than half a century, it also has the highest share of military expenditures relative to gross national income (7 percent compared to a world average of 2.3 percent).

Insufficient Growth in an Increasingly Competitive Environment

Growth of GDP in the MENA region averaged 3.2 percent a year over the last decade. However, countries and parts of the region differ—growth was 2.3 percent a year in the Maghreb and 2 percent in the Mashreq. Population growth rates are high, and growth in per capita income was sluggish at 1 percent for the MENA region as a whole. Per capita income was only 0.3 percent in the Maghreb, and even negative in the Mashreq where it fell by 1.6 percent annually with dire effects on unemployment and poverty.

By 2010, the labor force is expected to grow by more than 3 percent a year. This means that some 40 million additional jobs will be needed for young workers. Currently, unemployment rates in the MENA region average 20 percent of the labor force in economies outside the Gulf Cooperation Council (which includes Bahrain, Kuwait, Qatar, Saudi Arabia, Oman, and the United Arab Emirates) (Keller and Nabli, 2002). Today only one out of five new entrants finds a job. Polls in the region show that unemployment is the principal concern among young people (UNDP, 2002, p. 30).

MENA countries currently face a state of emergency. They suffer from the consequences of the global recession of 2001–2002 and the economic impact of September 11 (reduction in tourism revenues, difficulties related to airline security), as well as the ominous realities of drought and war.

At the same time, the global environment is increasingly uncertain and competitive. Global growth is recovering sluggishly and prospects remain clouded. Pessimistic forecasts in financial markets discourage recourse to capital markets, and MENA countries now face greater spreads. FDI in the region, which is already about half of the flows to emerging countries, is diminishing. Tourism and the income of emigrant workers, which together amount to about 10 percent of GDP (and 30 percent in countries such as Jordan), are sources of further uncertainties.

The economic environment will become increasingly competitive because of the establishment of a free-trade area with the European Union (EU), the phasing out of the Multifiber Agreement, membership in the EU of eastern European countries, the accession of China to the World Trade Organization (WTO), and trade liberalization among countries of the region through the Great Arab Free Trade Area (GAFTA), the Agadir agreements, and bilateral agreements.[1]

The Need to Change the Development Model

The situation must change to allow for greater per capita income in the long term and to create the necessary jobs. The aim should be growth of 6–7 percent a year, taking into account present salary levels and price fixing modes.

Saturated capital accumulation: The current growth model is rigid. Capital accumulation is significant (23 percent of GDP as compared to 30 percent of GDP in South East Asia) but has reached a ceiling. Growth of physical capital per worker became negative in the 1990s (Keller and Nabli,

---

1. GAFTA was established by the Cairo Convention of 1997. The regional free trade zone was inaugurated in 2001, with Jordan, Egypt, Tunisia, and Morocco as central actors. Also known as the "Agadir process," it aims to eliminate customs duties and commercial obstructions to multilateral trade in goods and services originating in the four countries (ahead of the 2010 target for the end of trade barriers in the Euro-Mediterranean area).

Figure 2.1. *Productivity Comparisons in the Textile and Clothing Sector, MENA Countries* (value added in dollars per employee)

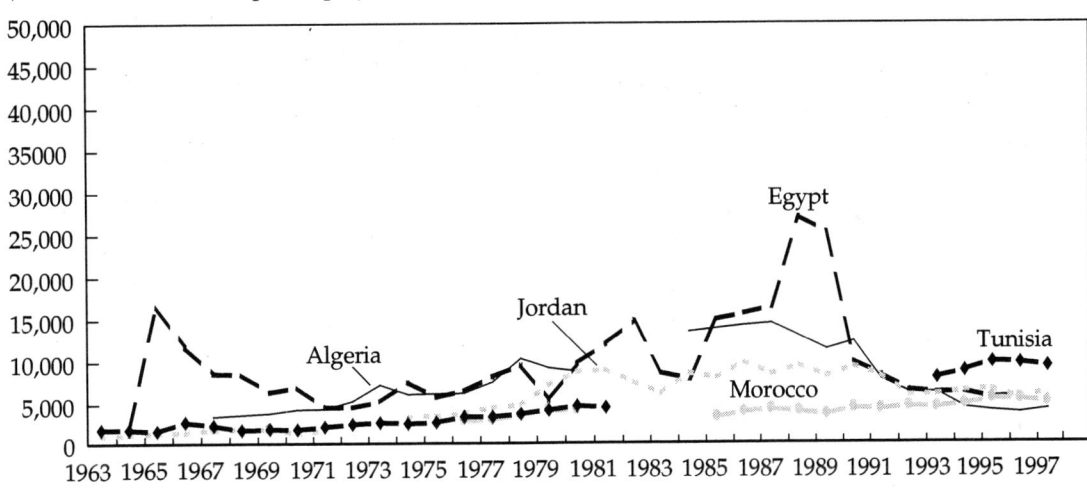

Source: Institut de la Méditerranée.

2002). In the past, the public sector was the main source of investment, but as public sector investment declined, private investment did not replace it. Capital accumulation suffers from malfunctioning capital markets and institutions. Moreover, there is too little FDI, which is usually more productive as it goes to more technology-intensive sectors or downward (service) industries.

Stagnant productivity: Overall productivity levels are low. Over the 1990s, the growth of GDP per worker was lower in the MENA region than in any other world region, averaging only 0.8 percent a year (Keller and Nabli, 2002). Labor productivity has suffered from low skill levels and poor organization of the labor force. The stagnation of labor productivity, together with higher salaries, has raised the cost per unit of labor for traditional export products such as foodstuffs, textiles, clothing and leather, and small mechanical products. Figures 2.1 and 2.2 compare productivity in MENA countries' textile and clothing sector with that of certain competitors. Their higher unit labor costs are leading to a significant loss of competitiveness in the main export markets and in the region's internal market.

Lack of diversification: Because they are insufficiently diversified, MENA economies have often proved unable to absorb internal shocks (such as rain levels) as well as external shocks. The economy requires a much more diversified and flexible basis of production.

High unemployment: In MENA economies, technological progress is slow, capital accumulation has reached a ceiling, and there is high unemployment. To maintain employment levels, labor productivity is kept low. In the absence of steady and continuous technological progress, countries restrain substitution of capital for labor in order to stabilize employment.[2]

---

2. The point can be formalized as follows (Blanchard, 1997, pp. 511–514):

$Y = F(K, AN)$ takes Y = production, K = the capital stock, A = state of technology, N = labor.

If capital is left out to simplify matters, $Y = AN$: the state of technology; A also stands for labor productivity. Whenever productivity (Y/N) rises, employment (Y/A) slumps. We therefore have the following relation: percent of employment variations = percent of product variations − percent of labor productivity variations. In the case of the MENA region, where the product growth rate is 3 percent, labor productivity must be brought down nearly to zero to absorb the 3 percent growth in the (constantly unemployed) labor force.

Figure 2.2. *Productivity Comparisons in the Textile and Clothing Sector, Other Countries*
(value added in dollars per employee)

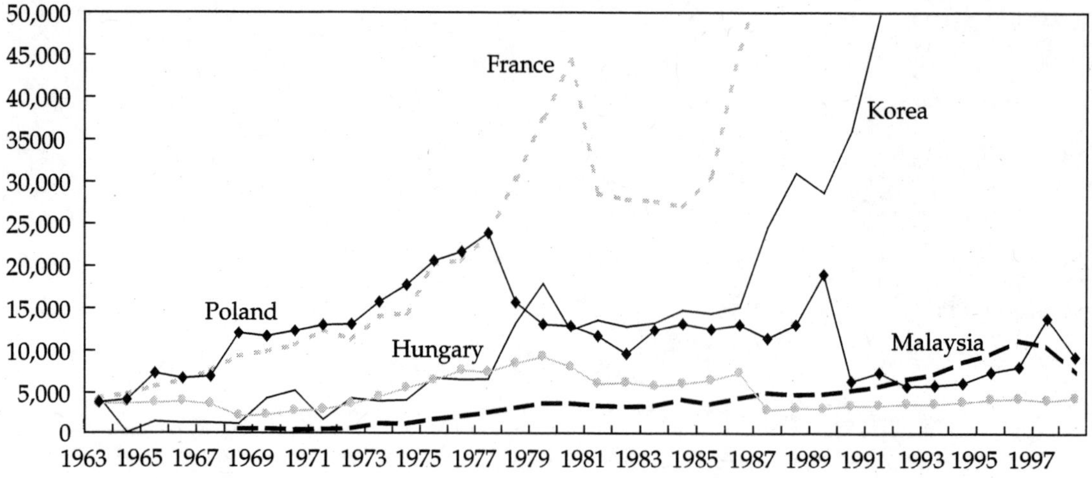

Source: Institut de la Méditerranée.

A scenario for development would include increasing the efficiency of the present system of growth through accumulation. This implies a set of measures to improve the overall economic framework to liberalize markets, notably the capital market, to adopt an active competition policy, to modify wages and price fixing systems, and to rationalize or privatize to reduce the level of structural unemployment. In addition, it would be necessary to undertake a resolute and determined transition from a growth model essentially based on accumulation to one founded on technological progress, innovation, and continuous learning. This makes the difference in growth performances among countries and is reflected in total factor productivity (TFP) (box 2.1).

Although the precise nature of TFP needs to be better understood, it is admitted to be a process that differs from the normal accumulation process and is based on externalities and increasing returns. It is at the core of the knowledge-based economies. Chapter 3 discusses in greater detail current developments in knowledge-based economies and related policy concepts.

Box 2.1. *Why Growth Performances Differ—Total Factor Productivity*

Differences in GDP per capita growth rates are explained less by differences in capital accumulation—both physical and human—than by total factor productivity (TFP). Recent empirical work by the World Bank, based on production function estimates for a large sample of countries, shows that most of the growth rate is to be attributed to TFP. The effect increases if improvement of human capital is added.

Variance decomposition: Contribution of	*TFP growth*	*Capital growth*
Without human capital (60 countries, nonoil exporters)		
1960–1992	0.58	0.41
1980–1992	0.65	0.21
With human capital		
1960–1992 (44 countries)	0.94	0.52
1980–1987 (50 countries)	0.68	0.20

Source: Easterly and Levine, 2001.

3

Knowledge and Economic Development: Recent Trends

Knowledge and innovation—the concrete application of knowledge in the form of new and improved technologies—has always been the driving force behind the development of societies. However, these have taken a quantitative jump over the past decade in the wake of the "explosion" of information and telecommunication technologies, the globalization process, and dramatic advances in the life, materials, and energy sciences.

These developments have led to new industries and new services, as well as to the renewal of established ones. Countries' competitiveness and welfare depend more than ever on their ability to create and use knowledge throughout the economy. The implications of the emerging knowledge-based or knowledge economy for the way economies function and governments should envisage their strategy are considerable.

The Knowledge-Based Economy

The knowledge-based economy has a certain number of characteristics (Lundvall, 1998):

- Innovation is a permanent feature. The rapid rhythm of change differentiates it from previous technological revolutions, which also showed a marked recourse to new information.
- It is an economy of networks at different hierarchical levels. Global networks dominate the top of the pyramid, and a growing number of excluded entities (which in one way or another also constitute networks) lie at the bottom.
- It is accompanied by new forms of organization involving industrial cooperation, polarization, and relations between the public and private sectors.
- Human capital plays a decisive role, and the capacity to learn matters more than the level of knowledge. While secondary school certificates were the trump cards of industrialization, higher degrees are those of the knowledge economy. Lifelong training is essential.
- Tacit knowledge needs to be codified and distributed.
- Information-related activities proliferate in all sectors of the economy.
- Finally, it challenges traditional economic theory (box 3.1).

Box 3.1. *Challenges to Economic Theory*

In the knowledge-based economy, growth is based on learning, which facilitates the endogenous development of resources and implies that the stock of knowledge matters less than its renewal. Knowledge generates externalities and rising outputs, and the market functions according to asymmetrical information. Knowledge is an unlimited and renewable resource but it is difficult to appraise because tacit knowledge is as important as formal knowledge. Ownership of knowledge is hard to establish since it is nonexclusive. It is a nonrival good since its use by one person does not exclude its use by others. Depending on its degree of codification, it can be ascribed to a physical person.

Viewed from a more sectoral angle, the knowledge economy develops high-technology in-dustries, particularly in ICT and services, which are the main job providers. This does not mean that all jobs require high qualifications. Nonetheless, the role played by tangible capital decreases in favor of the intangible capital constituted by the education and training of the labor force and the applied knowledge acquired though domestic R&D or by tapping into the global stock of knowledge. Organizational structures and practices change, with more decentralized manage-rial structures functioning as networks, systems of information, monitoring processes, market-ing, and interfaces connecting users and clients.

Industrial economies, whose characteristics are summarized in box 3.2, have endeavored to adapt to this new era of economic development. Some countries have moved more rapidly than others. English-speaking and Nordic economies and Korea have shown more receptiveness to new technologies and more flexibility in their overall institutional and economic frameworks.

It is not only the most advanced economies that have taken advantage of knowledge economy trends. Ireland used to be among the less developed Organisation for Economic Co-operation and Development (OECD) economies, but it is now thriving and may become a center of European

---

Box 3.2. *Trends in Knowledge and Innovation in the OECD Area*

Investment: At about 8 percent of GDP, knowledge-related investment (comprising R&D, software, and public education) is equal to investment in equipment. It has grown much faster than GDP and has tended to replace classical investment in equipment.

Learning: The great majority of active adults have reached at least the primary and lower secondary levels of education. However, differences among countries are greater for the upper secondary cycle and higher education (in the OECD area as a whole, over 14 percent of the labor force has a univer-sity degree).

Research and development: OECD countries allocate an average 2.2 percent of GDP to R&D. Countries that are rapidly "catching up" (Korea, Ireland) have relatively high percentages (2.8 percent and 1.5 percent, respectively). The business sector tends to represent an increasing share of R&D (50 per-cent and more in advanced economies).

Information and communications technology: Spending on ICT rose sharply during the mid-1990s in the OECD area to more than 6 percent of GDP and more than 8 percent in the countries most actively engaged in such technologies (Australia, New Zealand, Sweden, the United States). The OECD area also has by far the greatest share of computers and over 90 percent of Internet access. As Internet penetration is closely linked to cost, the countries in which the cost of Internet access is low are also those with the most developed services.

Innovation: In most OECD countries, "innovative" firms (generally defined as those that have intro-duced new technologies or improved processes within the last five years) range from 60 percent to 80 percent of all firms (assessed on the basis of size) and are equally involved in both industry and services. On the basis of these criteria, countries at the top of the pyramid also have the most publica-tions per inhabitant, the most patent applications, and the most firms involved in international coop-eration in R&D.

Productivity: Finally, these countries had labor productivity growth rates of 2–4 percent a year at the end of the 1990s, despite a relatively high level at the outset. Labor productivity in the United States is among the highest in the world and reached an average annual growth rate of 2 percent between 1985 and 1997 with an acceleration at the end of the period that is undoubtedly due to the develop-ment of its knowledge-based economy.

high-technology production. On the other side of the world, in India, regions such as Bangalore and Hyderabad have a very rapidly growing software industry of worldwide significance. In South America, Chile, the best-performing economy, has made extensive use of knowledge and innovation to develop comparative advantages in natural resources and agriculture (boxes 3.4–3.6 at the end of the chapter).

Key Pillars of Knowledge-Based Economies

The conceptual framework designed and systematically applied by the World Bank Institute to assess knowledge-based development strategies in different regions of the world indicates that, to develop, a knowledge-based economy requires the following four "pillars":

- An economic and institutional model that provides incentives for the efficient creation, dissemination, and use of knowledge to promote growth and increase welfare;
- An educated and skilled population that can create and use knowledge;
- An innovation system composed of firms, research centers, universities, consultants, and other organizations that can tap into the growing stock of global knowledge, adapt it to local needs, and transform it into products valued by markets;
- A dynamic information infrastructure that can facilitate the effective communication, dissemination, and processing of information.

A fifth pillar addresses the intangible factors that make a society function efficiently and move forward, such as the capacity to formulate a vision, the level of trust and self-confidence, and the appropriateness of guiding values. In fact, these qualitative elements are the driving force in the move toward new models of development.

The Knowledge Economy and Economic Performance

To assess countries' readiness for the knowledge economy, countries must be positioned in relation to others through the use of appropriate indicators. To this end the World Bank Institute has developed a database covering 100 countries and 69 variables calculated from raw data for the period 1995–2000. This material is then computed into indexes that reflect a country's performance with respect to each pillar of the knowledge economy. These indexes compare one country to others in relative rather than absolute terms. A selection of 12 variables (box 3.3) gives an

---

Box 3.3. *Variables Selected for Knowledge Economy Benchmarking*

For each of the knowledge economy pillars, three variables were selected:

- The economic and institutional pillar: tariff and nontariff barriers, property rights, government regulation;
- The innovation pillar: number of researchers in R&D, share of manufactured products trade in GDP, number of scientific and technical publications per million inhabitants;
- The education pillar: literacy rate, secondary school enrollment, higher education enrollment;
- The infrastructure pillar: telephone lines, computers, Internet access, all per thousand population.

In order to position countries on a comparative scale from 1 to 10, variables are normalized. The normalized value of a variable u is calculated as follows:

$$\text{Normalized }(u) = 10*\{\max[\text{rank}(u)] - \text{rank}(u)\}/\{\max[\text{rank}(u)] - 1\}$$

overall view of the main World Bank Regions and countries' knowledge economy readiness. An overall knowledge economy index is computed as the average of the scores obtained for each of the 12 (normalized) variables.

The Knowledge Economy and GDP

There is a strong correlation between a country's overall knowledge economy readiness index and its level of development as measured by GDP per capita (figure 3.1). This can be interpreted in two ways. On the one hand, investment in factors related to the knowledge economy in past decades contribute decisively to growth performances (see the data on TFP in box 2.1). On the other hand, the higher the level of development, the greater the ability to invest in aspects of the knowledge economy and establish an appropriate economic and institutional framework to take advantage of them. This is of concern to the extent that the gap in development capabilities is likely to widen in parallel to the "knowledge gap," accelerating the process of divergence among countries (Easterly and Levine, 2001).

Knowledge Economy and Competitiveness

By ranking countries on the overall knowledge economy index it is possible to compare their competitiveness. The World Economic Forum provides a prospective "growth competitiveness" index that uses variables characterizing technological performance, institutional appropriateness, and macroeconomic stability. There is also a strong correlation between the ranking for this index and the ranking for the knowledge-economy index (figure 3.2).[1]

Figure 3.1. *The Knowledge Economy and GDP per Capita*

GDP per Capita

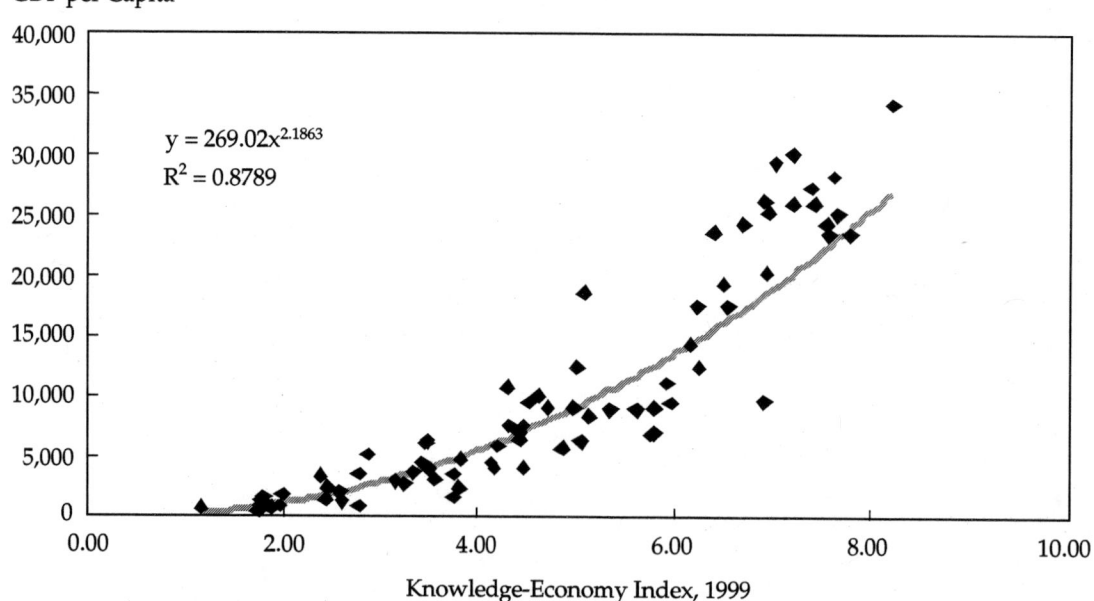

Knowledge-Economy Index, 1999

Source: World Bank Institute.

---

1. This can be partly explained by some overlap of the variables used in the two indexes. For details on the methodology used in the World Economic Forum Competitiveness Index see World Economic Forum (2002).

Figure 3.2. *Knowledge-Economy and Competitiveness Ranking*

Growth-Competitiveness Index (GCR) Ranking

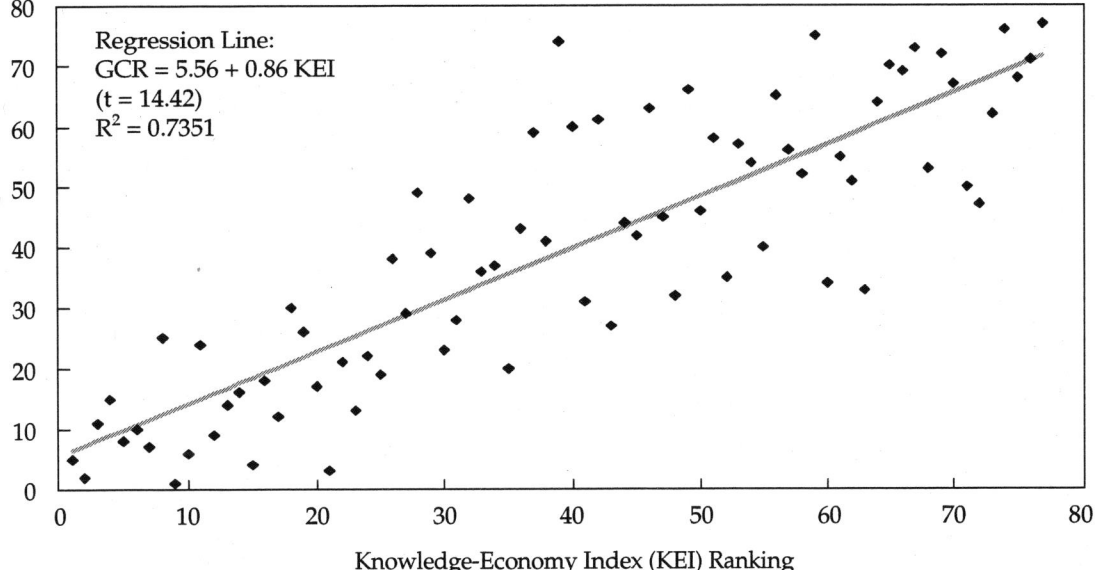

Regression Line:
GCR = 5.56 + 0.86 KEI
(t = 14.42)
R^2 = 0.7351

Knowledge-Economy Index (KEI) Ranking

Source: World Bank Institute.

Guidelines and Lessons from International Experience

First Guideline: Developing a Holistic Strategy

Countries must take a systemic approach encompassing the four basic pillars. This is clear from the fact that applied studies have been unable to isolate the impact of any specific knowledge economy component (pillar) on growth. In particular, the effect of education and training, estimated on the basis of schooling and educational expenses, is not clear, although cross-country analysis of a large number of countries reveals a positive relation between human capital and growth (for details, see Boutrolle, 2002). Educational criteria (and other knowledge economy components) are therefore on a par with levels of wealth or growth rates, hospital beds, or highway miles. This suggests that single knowledge economy components will have a weaker "edge" than the full complement of relevant components. Thus, when these components are combined into a larger and more organized system, they have a decisive impact on growth.

Second Guideline: Generalizing the Knowledge Economy to All Sectors

It would be naïve to hope to create a knowledge economy by massively financing organizations or sectors responsible for the production and transmission of knowledge. To have a maximum effect on overall productivity and to bridge the gap between the various networks within a society, the knowledge economy requires going well beyond the development of one or several specialized sectors (Eliasson, 1996). The production and processing of knowledge are part of all economic activities, including traditional or basic activities of low technological intensity. The necessary system not only ensures continuous growth in specialized sectors but also allows for the generalization of knowledge-intensive activities throughout all sectors of the economy.

This makes it possible to understand why the most advanced economies were able to have growing labor productivity alongside diminishing unemployment rates. While the introduction of the knowledge economy did lead to some extent to the substitution of capital for labor at the expense of employment, workers were rehired as overall productivity increased.

Third Guideline: Accelerating Structural Reforms

To become a knowledge economy, the whole economic and social fabric must adapt quickly and in depth. This makes structural reforms of all types—in finance, trade, the labor market, and so on—even more important than in the past. Such structural reforms create a climate that can facilitate the passage to a knowledge economy:

- A competitive business framework creates incentives that encourage and reward innovation.
- Financial sector reforms can help develop the nonbank financial sector, which plays a central role in financing innovative businesses.
- Policies that promote trade and investment encourage business innovation because of external competitive pressures and knowledge transfers in terms of best business practices.
- More flexible labor markets facilitate mobility and the employment of skilled personnel in the most dynamic firms.

Fourth Guideline: Adapting Investment for Learning to Levels of Economic Development

Countries at every level of development can become knowledge-based economies in some way. However in no case can investment in cognitive fundamentals be bypassed. Basic reading, writing, and calculating capabilities, and basic literacy more generally, are essential. These should be complemented by the acquisition of "functional literacy", which allows people to use their knowledge in autonomous and creative ways to "navigate" daily life and their professional milieu.[2]

Qualification levels must be constantly adjusted as the economy climbs the value chains in global production and the division of labor. At the early stage, when the country depends largely on FDI, basic technical skills are acquired through vocational and technical education. As industries become more sophisticated and are able to innovate, more investment in higher education is required and more students need to enroll (for a detailed analysis, see World Bank, 2002).

When developing an indigenous R&D capability, achieving a world-class level in any area generally requires huge investments and a long process of knowledge accumulation. In most fields, it is much more profitable to tap into the global stock of knowledge through mechanisms such as licenses, capital goods trade, or specific international technological intelligence instruments.

Adapting Knowledge Economy Strategies to Specific Countries' Characteristics

Knowledge economy strategies must be adapted to each country's needs, taking into consideration the structure of the economy, the level of development, and the country's socioinstitutional peculiarities. In a sense, it might be preferable to speak of a knowledge for development (K4D) strategy rather than a knowledge economy strategy, a notion that better suits developing countries' perspectives. Boxes 3.4, 3.5, and 3.6 illustrate how different economies have applied the policy and strategy principles sketched above.

---

2. Functional literacy is evaluated in surveys such as the OECD International Adult Literacy Survey (IALS).

Box 3.4. *Ireland, the Celtic Tiger*

Ireland has demonstrated that a country traditionally labeled one of the poorest members of the European Union, highly dependent on agriculture and low-end manufacturing, can successfully turn its economy into a provider of high-technology services. Ireland's transformation is attributable to sustained and well-targeted investment in education and to a policy framework favorable to FDI, notably in the ICT sector. At 20 percent of GDP it has one of the world's highest net inflows of FDI, second only to Sweden.

Ireland has become one of the most dynamic knowledge-based economies in Europe and is the second largest exporter of software. With an average rate of growth in GDP of 9.1 percent over the period 1995–2001, the "Irish miracle" is not attributable solely to the government's investment in education and its efforts to attract FDI. Ireland's success also stems from a stable macroeconomic and fiscal environment and significant openness to trade.

Substantial EU assistance has also helped Ireland to target investments relevant to a knowledge economy. Today, it is the headquarters of many European technology giants, and Dublin has taken advantage of its well-developed network infrastructure to become the hub for European telephone call centers. Ireland has thus come a long way from its traditional low-end manufacturing economy, but to become a full-fledged knowledge economy, it has to strengthen indigenous innovation.

Box 3.5. *India, from the Silicon Valley of South Asia to a Global Knowledge Superpower*

India does not rank very high in most indexes of technological progress. Its software industry, however, is recognized worldwide as a success. In 2000–2001 its software exports accounted for 14 percent of total exports with revenues of US$ 6.2 billion and a growth rate 55 percent above that of the previous year. This success is not only one of volume, but also of quality and technical excellence.

Bangalore and Hyderabad illustrate the cluster-based development process that took place. Bangalore's success is based on a business environment and a special legal framework that promotes scientific businesses. In addition, the state and central governments have tried to provide state-of-the-art facilities to attract and retain the most talented specialists. This shows that removal of a traditional planned economy bureaucracy releases the potential for economic growth.

Building on its regional successes, India has developed a strategy for becoming a knowledge superpower. Plans have been made to reduce illiteracy, to mobilize the broad, high-quality network of national technological and management institutes as a core training and research base, and to expand the ICT infrastructure, with a view to offering universal access with innovative, user-friendly technologies.

A New Mindset for Government Action

In a "knowledge economy" approach governments play a primary role in their function as orientating authority. They provide the essential economic and institutional framework and must encourage the necessary investments in knowledge, innovation, and new technologies with appropriate incentives. If a country is to become a knowledge economy, government action requires a new mindset. The role of government should not diminish but be adapted to the needs of this new economy.

In recent decades, development policies have been guided by ideas of liberalization and then of modernization. The guiding principle of liberalization was "undoing things." The government focused exclusively on economic issues and left agents free to act. The guiding principle of modernization was "building things." Government acted as a good regulator and focused on establishing modern institutions and a good basic environment, with more attention to the social

---

Box 3.6. *Chile—Knowledge, Innovation and New Comparative Advantages*

Chile is regarded today as the most developed economy in Latin America. It has enjoyed an average annual growth rate of 6.5 percent over the past decade and in 2003 the World Economic Forum ranked it 20th in growth competitiveness (a significant improvement from its 27th rank in 2002). A sound macroeconomic framework with bold reforms initiated in the 1980s and sustained investments in human resources have provided the basis for Chile's performance. A distinct feature of the Chilean strategy has been to exploit or create comparative advantages in natural resources. Particularly striking are the competitive positions acquired in fish farming and viticulture, areas in which Chile did not possess a particular advantage or specific know-how two decades ago. Making the best use of knowledge and technologies available worldwide, these industries have flourished and become leading exporters.

Instrumental to this success is a semipublic innovation agency called the Fundación Chile, which developed a remarkable set of instruments to accompany Chile's transition to a knowledge economy. International networks of experts tapped into foreign knowledge and know-how, efficient systems of skill qualifications and certifications for the labor force employed in those industries were established, and venture capital funds were strengthened. However, a full-fledged dynamic innovation system must be built to ensure Chile's long-term competitiveness.

Source: Introductory Note to the Forum on Knowledge for Development held in Marseilles.

---

dimension. In the knowledge economy, it is important to build winning opportunities. Here, the government should provide a long-term vision and integrate policies with a view to making the country globally competitive. These ideas are summarized in table 3.1. The knowledge economy framework does not reduce the relevance of the liberalization and modernization frameworks, notably in countries and regions (such as MENA) which have not yet made enough progress in these directions, but it adds new policy dimensions, with government playing a visionary and motivating role while promoting essential structural reforms.

Table 3.1. *Liberalization, Modernization, and Knowledge Economy Mindsets*

Mindset	Liberalization mindset	Modernization mindset	Knowledge economy mindset
Is About	Undoing things	Building things	Building winning opportunities

Creates	Freedom
	Fluidity
	Even playing field
		Modern institutions
		Rule of law
		Good basic business
		  environment
			Vision
			A winning mentality
			Clusters
			A vibrant home base for
			  business

Main focus	Stability, incentives	Productivity catchup	Becoming globally competitive
Domain	Economy	Economic, social	Societal
Role of government	Get out of the way Stop being an operator	Become an integrator	Become a challenger Become a good regulator

Source: Rischard, 2002.

4

MENA Countries' Readiness for the Knowledge Economy: A Snapshot

To assess countries' readiness for the knowledge economy, it is necessary to position them using appropriate indicators. To this end, the *Institut de la Méditerranée* developed a database covering 80 countries and 31 variables for the 1995–2001 period, some of which are from the World Bank Institute's knowledge assessment methodology database (www1.worldbank.org/gdln/kam.htm). Selected elements are listed in chapter 3, box 3.3. These variables are used in a multicriteria analysis to reveal countries' performance with respect to each pillar of the knowledge economy (box 4.1).

The appendix presents a more detailed analysis and also gives a dynamic country-by-country picture over the period 1995–2001 (or most recent year).

Overall Knowledge Economy Readiness

Figure 4.1 shows that the region's overall knowledge economy readiness is at the low end of Class 3. Only South Asia and Sub-Saharan Africa rank lower. The MENA region[1] experienced a slight deterioration over the period 1995–2001 and therefore did not close the gap with East Asia (little deterioration) or Eastern Europe and Central Asia and Latin America, which improved their performance (see the appendix).

As a whole, the MENA countries' overall knowledge economy position is not in line with their level of economic development (figure 4.2). As will appear, the overall ranking is mainly explained by weakness in the MENA region's incentive and institutional framework (pillar 1: middle of Class 2). In the three other pillars, the region is very close to the global average (lower middle of Class 3). This suggests that weak incentive and institutional mechanisms play a determining role in the region's overall knowledge economy performance, as a closer look at countries like Saudi Arabia and Kuwait shows. These countries stand at the top for a number of educational variables (such as expenditures on education) and for development of the new economy (Internet users). However, their mediocre performance in the first pillar variables (except black market control and control of corruption) negatively affects their overall knowledge economy position. Jordan also shows the importance of pillar 1, as it appears to be particularly well positioned in terms of the knowledge economy, despite its lower per capita income.

Clearly, the region's considerable effort to ensure "universal schooling" has not been sufficient to spur growth and employment and encourage the development of a knowledge economy. The inappropriateness of the efforts that have been made may explain this worrisome situation.

The region has experienced a relative deterioration in the ICT infrastructure pillar. The total number of telephones per inhabitant has risen sharply, mainly because of mobile subscriptions. The relative position of fixed lines has worsened and affects growth of the Internet, especially for high-speed access. The lack of computer skills and the negative impact of information control explain the relative deterioration of the region's position in terms of Internet users.

---

1. Note that Israel is included in the MENA region, based on an arithmetic average of the countries included in the World Bank Institute database. The regional groupings correspond to those defined and used by the World Bank.

Box 4.1. *Multicriteria Analysis and Variables Selected for Knowledge Economy Benchmarking*

The 31 variables (see below) give an overall view of World Bank Regions and countries. The ranking gives a picture of countries' knowledge economy performance in relative rather than absolute terms. Changes in countries' positions are also to be seen in relative terms, so that even if a country's absolute situation improves, its ranking may drop if other countries improve more.

The MENA sample includes the following ten countries: Algeria, Egypt, Iran, Jordan, Kuwait, Morocco, Saudi Arabia, Syria, Tunisia, and Yemen. For purposes of comparison, Turkey is also included. These countries' relative evolution over the 1995–2001 period can been found on individual country graphs in the appendix.

The variables for the various knowledge economy pillars are listed below and are indicated in later graphs by the number assigned to each.

- The economic and institutional pillar: 1: tariff and nontariff barriers; 2: freedom to use alternative currencies; 3: property rights; 4: freedom of capital and financial exchange; 5: regulation of FDI; 6: regulation; 7: black market control; 8: voice and accountability; 9: participation of women in labor force; 10: control of corruption.
- The innovation pillar: 11: researchers per 10,000 inhabitants; 12: FDI per 100 inhabitants; 13: trade (exports + imports) per 100 inhabitants; 14: science and engineering students (percentage of total students); 15: credit to private sector (percentage of domestic credit); 16: domestic credit provided by banking sector (percentage of GDP); 17: stocks trade turnover ratio (%); 18: market capitalization of listed companies (percentage of GDP).
- The education pillar: 19: primary pupil-teacher ratio; 20: public spending on education (percentage of GDP); 21: years of education; 22: adult literacy rate (percentage aged 15 and over); 23: secondary gross enrollment ratio; 24: tertiary gross enrollment ratio (male and female); 25: tertiary gross enrollment ratio (female).
- The ICT infrastructure pillar: 26: telephones per 1,000 inhabitants; 27: computers per 1,000 inhabitants; 28: Internet hosts per 10,000 inhabitants; 29: radios per 1,000 inhabitants; 30: daily newspapers per 1,000 inhabitants; 31: loss in electricity distribution (percentage of output).

The method involves assigning each country to defined classes according to a set of variables chosen to capture the logic of the four knowledge economy pillars. Each class is defined by a hierarchically organized norm (profile), and a country's assignment to a given class depends on its performance in a set (or subset) of indicators related to the norm. To be assigned to a class, a country's performance should be "at least as good as" the norm from the point of view of the chosen indicators. In the Institut de la Méditerranée ranking, presented in the following pages, there are five classes. Class 5 contains the countries that perform best in terms of the knowledge economy. According to the criteria chosen, they are in a better position than 80 percent of the world (the sample). Class 4 contains countries whose performance is weaker that those in Class 5, but perform better than 60 percent of the other countries. Countries in Class 3 perform less well than those in Classes 5 and 4, but better than 40 percent of the other countries. And so on.

In the innovation pillar, the MENA region also shows a relative deterioration. One explanation is the low level of scientific culture and a strong preference for the humanities over the hard sciences, which is aggravated by a marked brain drain. Another reason appears to be an increasing lack of risk capital and finance for innovation.

The countries of the region show a marked degree of dispersion. In fact, the MENA region shows the greatest dispersion of all World Bank Regions covered in the benchmarking effort. On the basis of the knowledge economy index, it is possible to divide the region into three groups. Jordan and Kuwait are above the MENA average. However, while Jordan improved significantly, Kuwait stagnated. A second, more heterogeneous group is composed of Tunisia, Morocco, and

Figure 4.1. *Knowledge Economy Readiness: MENA Countries*

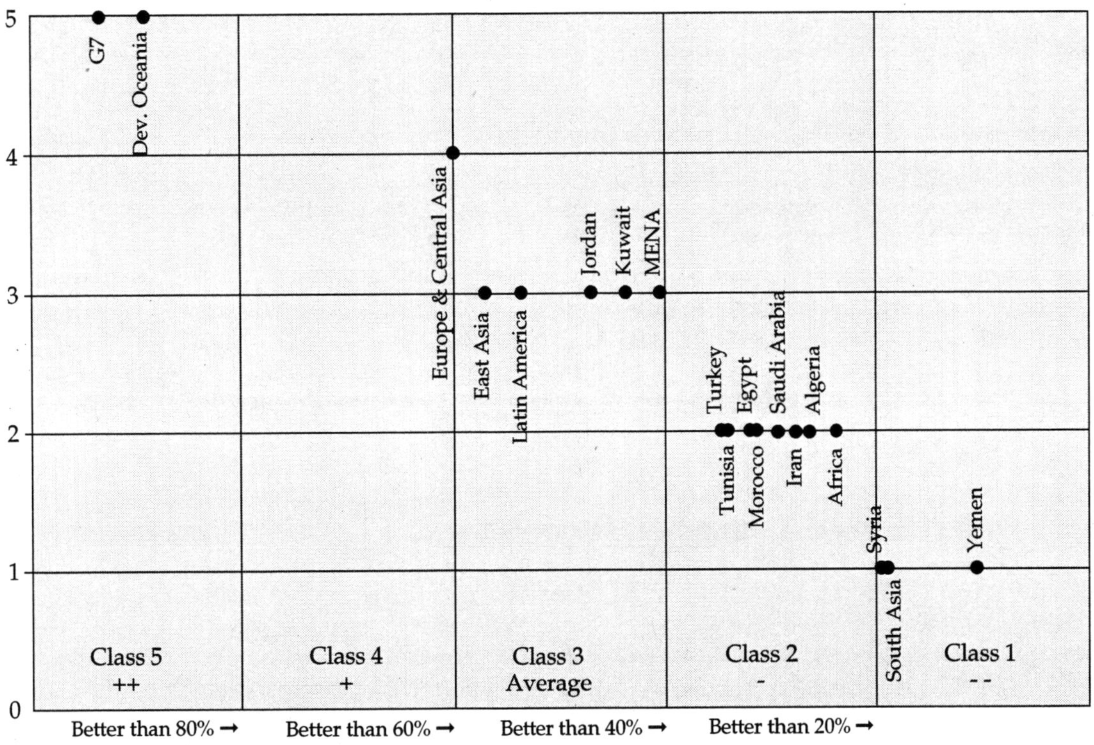

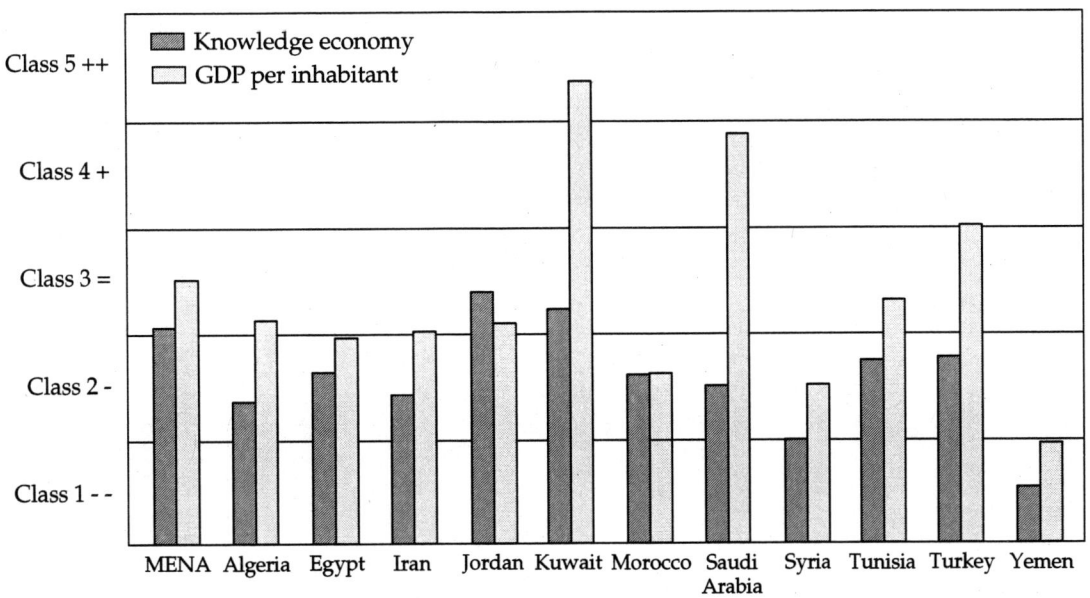

Source: Institut de la Méditerranée.

Figure 4.2. *Countries' Relative Position in the Knowledge Economy and GDP Per Inhabitant*

Source: World Bank, WDI, and Institut de la Méditerranée.

Egypt, followed by Saudi Arabia, Iran, and Algeria, which fall below the MENA average but are better positioned than Africa and South Asia. Syria and Yemen, which are at a very low level on the knowledge economy readiness scale, constitute the third group.

As Saudi Arabia shows, the best-positioned countries—according to the selected indicators—are not those endowed with large oil resources. The latter countries have not diversified output and investment or developed a strong financial system and an attractive institutional framework. While these were not essential in an oil-driven economy, their absence creates problems for a knowledge-driven one. Jordan stands out as a result of its dedication to a knowledge economy vision and its efforts to support it with the required investments (see chapter 9).

Algeria, Egypt, Morocco, and Iran have improved their ranking, but not enough to reach the MENA average. Iran's results on the various pillars differ: economic incentives remains insufficient, while reforms and investments in ICT and education have been significant.

The following sections analyze the region's performance in terms of the four components of the knowledge-based economy: economic and institutional framework, education and training, ICT infrastructure, and innovation.

Economic Incentives and Institutional Framework

Since 1995, the position of most MENA countries relative to the rest of the world has deteriorated (figure 4.3). Despite some improvements, especially in tariff and nontariff barriers to trade and control of corruption, most MENA countries continue to lag behind most other countries in this area.

As figure 4.4 shows, the region is weakest in terms of freedom to use foreign currencies and financial exchange, accountability of the legal framework, and participation of women in the

Figure 4.3. *Economic Incentives and Institutional Framework: MENA Countries*

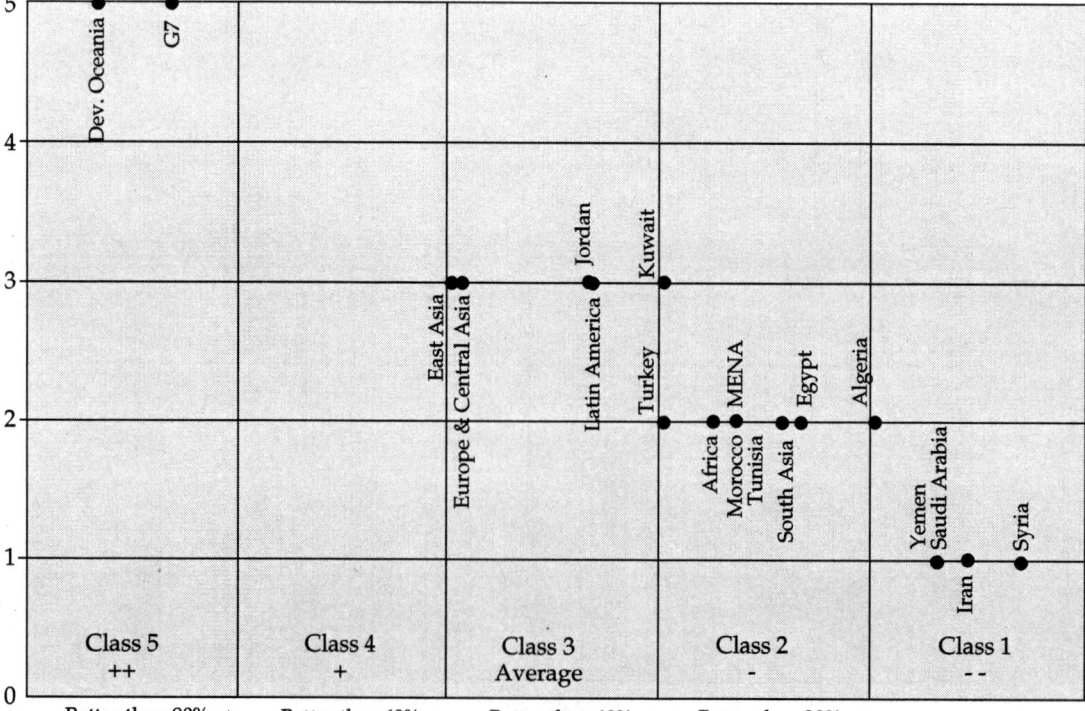

Source: Institut de la Méditerranée.

Figure 4.4. *Economic Incentives and Institutional Framework: The MENA Region's Relative Performance*

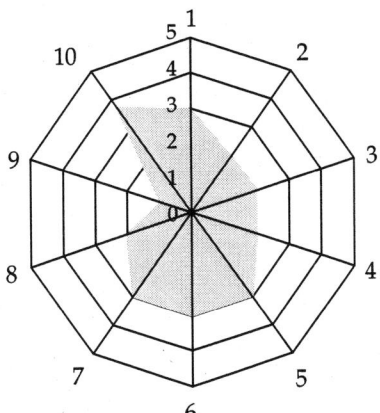

The economic and institutional pillar: 1: tariff and nontariff barriers; 2: freedom to use alternative currencies; 3: property rights; 4: freedom of capital and financial exchange; 5: regulation of FDI; 6: regulation; 7: black market control; 8: voice and accountability; 9: participation of women in labor force; 10: control of corruption.

Source: Institut de la Méditerranée database or most recent data.

labor force. As these are central conditions for investment and other efforts directly related to knowledge, information, and innovation, the potential payoff of knowledge-related investments in economic and social results is affected.

Despite their low initial position, most of the countries (except Jordan and Kuwait), have not improved their relative position in this pillar. Morocco, Syria, and Tunisia, in particular, have even experienced a significant deterioration, while Algeria and Egypt show some improvement.

Education

MENA countries formerly spent quite heavily on education (as a percentage of GDP), and they perform relatively well for the indicators measuring efforts on education: public spending, secondary school enrollment, and higher education enrollment (figure 4.5). However, these indexes have been decreasing and are now well below the world average (figure 4.6). Algeria, Yemen, and particularly Jordan show major improvements in their positions. The positions of Kuwait and Morocco have weakened.

Several issues need to be tackled. Illiteracy rates remain high, particularly among women. Difficulties in achieving universal education have meant a relatively smaller average number of years at primary, secondary, and high schools. The short duration of schooling results in an insufficiently qualified workforce.

Moreover, in most MENA countries unemployment has risen or employment has stagnated. This indicates the difficulty of dealing with a rapidly growing population and the inadequacy or lack of coherence between employment and education policies in most MENA countries.

Information and Communications Technology Infrastructure

In recent years several MENA countries have invested massively in ICT infrastructure. The gap between the MENA region and the rest of the world has decreased and a number of countries have moved up or even caught up with other regions (figures 4.7 and 4.8). In some countries—notably the Gulf States—the level of equipment is high. Telephone equipment per capita has increased significantly as a result of the development of mobile telephony. However, fixed lines have tended to increase slowly, with negative consequences for the Internet, which is constrained by bandwidth

Figure 4.5. *Education: MENA Countries*

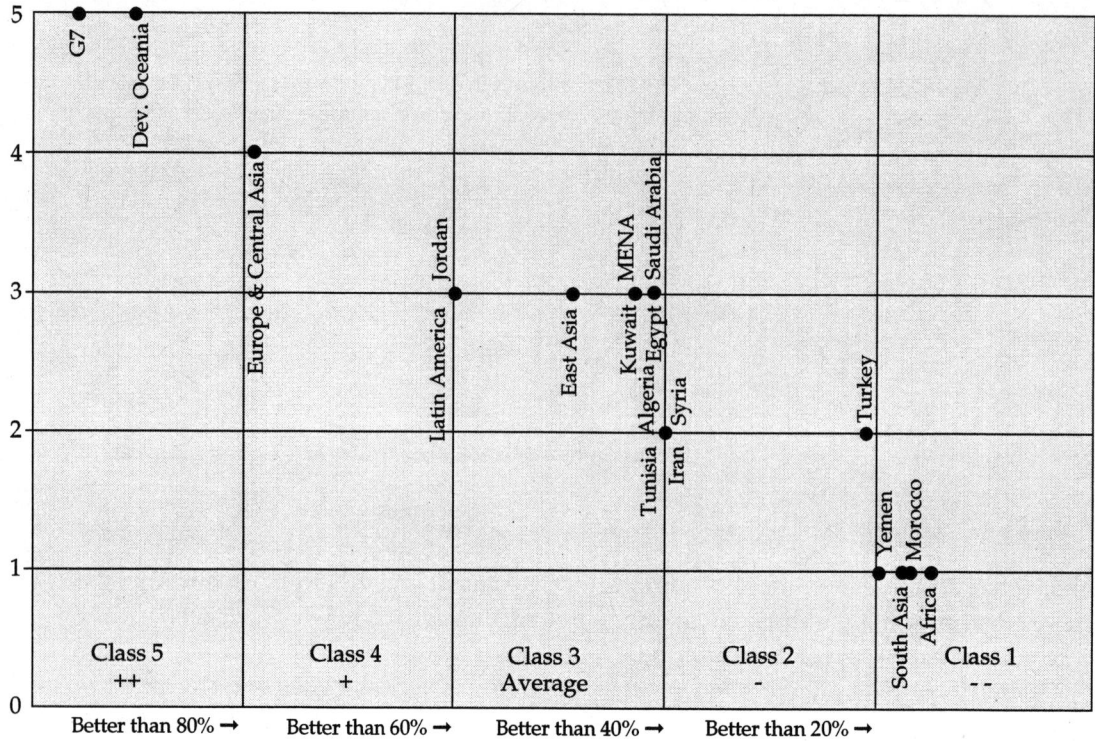

Source: Institut de la Méditerranée.

Figure 4.6. *Education: The MENA Region's Relative Performance*

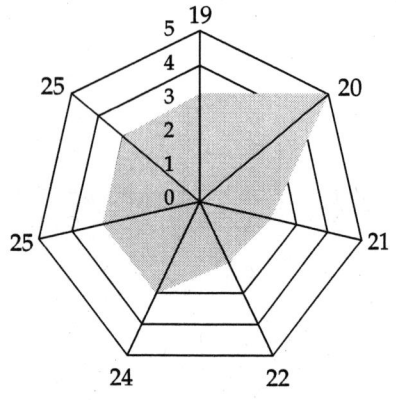

The education pillar: 19: primary pupil-teacher ratio; 20: public spending on education (percentage of GDP); 21: years of education; 22: adult literacy rate (percentage aged 15 and over); 23: secondary gross enrollment ratio; 24: tertiary gross enrollment ratio (male and female); 25: tertiary gross enrollment ratio (female).

Source: Institut de la Méditerranée database or most recent data.

capacity. The number of Internet users is advancing, but at a slower pace than the world average, resulting in a degradation of the region's relative position (see chapter 8 for further details).

Innovation

MENA countries show great dispersion on this index. Some countries (notably the small Gulf States) rank quite high, relative to the global position, while others have a much lower rank (figure 4.9). The "distance" between the G7 countries and MENA as a whole is the same as the

Figure 4.7. *ICT Infrastructure: MENA Countries*

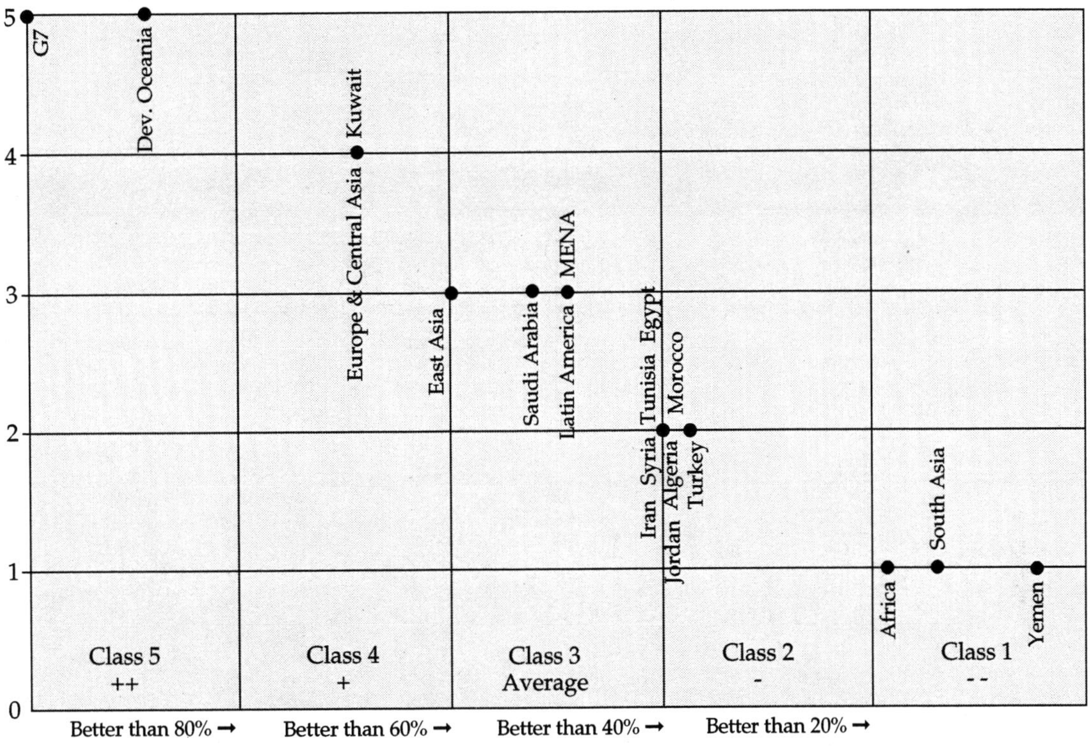

[chart: vertical axis 0–5, horizontal axis divided into Class 5 ++ (Better than 80%→), Class 4 + (Better than 60%→), Class 3 Average (Better than 40%→), Class 2 – (Better than 20%→), Class 1 – –]

Data points:
- G7 — 5
- Dev. Oceania — 5
- Europe & Central Asia — ~4
- Kuwait — 4
- East Asia — 3
- Saudi Arabia — 3
- Latin America — 3
- MENA — 3
- Iran, Syria, Tunisia, Egypt, Jordan, Algeria, Morocco, Turkey — ~2
- Africa — 1
- South Asia — 1
- Yemen — 1

Source: Institut de la Méditerranée.

Figure 4.8. *ICT: The MENA Region's Relative Performance*

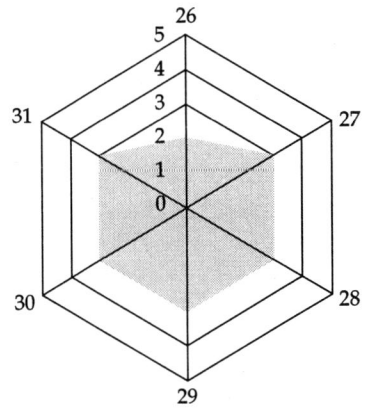

The ICT infrastructure pillar: 26: telephones per 1,000 inhabitants; 27: computers per 1,000 inhabitants; 28: Internet hosts per 10,000 inhabitants; 29: radios per 1,000 inhabitants; 30: daily newspapers per 1,000 inhabitants; 31: loss in electricity distribution (percentage of output).

Source: Institut de la Méditerranée database or most recent data.

distance between Jordan and Algeria. However, as figure 4.10 shows, the major problems relate to the institutional framework. The lack of investment, especially foreign investment, poor private sector access to domestic credit, and the weak liquidity of local stock markets clearly indicate that improvement of the financing of innovation should be a priority along with improvement of the institutional and regulatory framework.

The other main weakness of the innovation system is the content of the education system. Despite improvements in the gross enrollment ratio, the education system has been unable to overcome illiteracy and to train enough engineers and scientists.

Figure 4.9. *Innovation: MENA Countries*

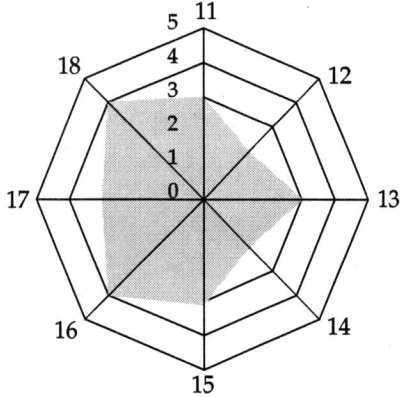

Source: Institut de la Méditerranée.

Figure 4.10. *Innovation: The MENA Region's Relative Performance*

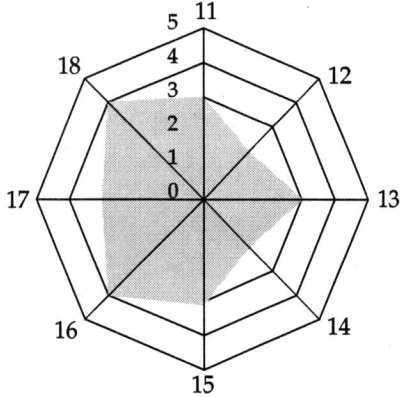

The innovation pillar: 11: researchers per 10,000 inhabitants; 12: FDI per 100 inhabitants; 13: trade (exports + imports) per 100 inhabitants; 14: science and engineering students (percentage of total students); 15: credit to private sector (percentage of domestic credit); 16: domestic credit provided by banking sector (percentage of GDP); 17: stocks trade turnover ratio (%); 18: market capitalization of listed companies (percentage of GDP).

Source: Institut de la Méditerranée database or most recent data.

The picture presented by the innovation pillar illustrates the interdependence of the four pillars. It is quite easy to look at the pillars separately, but it is also clear that a knowledge economy strategy cannot rely on disparate reforms in certain fields. It requires a systemic approach which encompasses all pillars and a broader shift in the society and related mindsets and behaviors.

5

Social and Economic Frameworks

A knowledge economy cannot take shape in the absence of a proper institutional and economic framework. To a large extent, what is at stake is countries' adaptation to the challenges of the 21st century.[1] Fundamental attitudes to knowledge and innovation need to evolve to improve the overall receptiveness to change. This chapter draws attention to issues that need to be addressed but does not discuss them in depth.

Power and Social Structures

Political Governance

Freedom in all forms is essential for developing a dynamic economy. The recent UNDP *Arab Human Development Report* (2002) mentions the deficit of political freedom as a preeminent issue and draws attention to noteworthy results from international surveys, notably those of the World Bank. Particularly problematic are the questions of freedom of expression of citizens and of government accountability. As the UNDP report suggests, various forms of participation need to increase: more democratic elections, greater involvement in the political debates of representative, independent nongovernmental organizations (NGOs), and free development of the media.

Endemic conflict in the region prevents progress and inevitably encourages authoritarian practices and controls. Some countries in the region have nonetheless been able to introduce significant reforms and shown the way forward. Developed countries and donors can promote the establishment of more open and democratic societies through appropriate initiatives, and change throughout the region can be accelerated by a sort of domino process in which one country follow the lead of another (Yacoubian, 2002).

Most MENA countries are also affected by strong information control, which takes many forms, notably various restraints on the use of the Internet (censorship, high prices, poor infrastructure). The control of information has been, and still is of course, an essential attribute of power and helps those who have it to keep it. But it is problematic for countries that aspire to a knowledge economy which is predicated on a rich, diverse and open exchange of all sorts of information. Many developing countries are now making use of e-government instruments to modernize public administrations in various areas (voting, taxes), and such instruments could be usefully applied in MENA countries (Saidi and Yared, 2002).

Modernization of the central government and reforms in the exercise of power are needed in many policy areas of crucial importance for the establishment of a knowledge economy, such as education and training systems and regulation of telecommunications (see chapters 6 and 8, respectively).

---

1. Some authors see the origin of the problematic development of the state in the Muslim world in the fact that there has been no constituted church as in the Christian world. Such a constituted church in effect functioned as a dominant and established locus of doctrine and power, with a dedicated corpus of personnel, to the end of the Middle Ages, preparing the ground for the state institutions in the form familiar to Western societies (Ruthven, 1997).

Status of Women

Women's low level of participation in political organizations and in the economy and their low literacy rate considerably affect the dynamism of a number of MENA countries. A human resource that is fundamental for the dynamism and creativity of a knowledge economy is being ignored.

Table 5.1 provides women-related indicators that make it possible to compare the MENA region to other world regions. Two indicators—the share of women in the labor force and the percentage of women enrolled in tertiary education—were introduced in the expanded database of 31 variables presented and discussed in chapter 4. Given the modest weight of these two variables (less than one-fifteenth), the impact on the overall benchmarking picture is limited. The introduction of these gender-related variables, however, allows Iran and Tunisia to gain a better relative ranking, while that of Saudi Arabia deteriorates.

Historical and sociocultural elements, such as regulations on inheritance, which precluded women from owning land, or the taboo on women associating with men outside the *mahram* (i.e., the prescribed group of men with which a Muslim woman is allowed to associate outside the home), has resulted in the exclusion of women from public life and are often used to justify the status of women. Such arguments are less cogent today. Tunisia, Egypt, and Lebanon have taken the lead and provide good examples of reform. A systemic approach is likely to be needed to tackle this problem in the areas of education (literacy, access to university studies), the legal framework (divorce rights), and the labor market (including the opportunities offered by teleworking for home-based jobs).

Investment Climate

Bureaucracy

Widespread bureaucratic practices that affect all aspects of entrepreneurship and investments—domestic as well as foreign—deserve special comment. Bureaucratic hurdles are prejudicial to emerging entrepreneurs and small firms. Enterprise surveys by the World Bank and other organizations show that this is a very serious problem in more than one MENA country (box 5.1). Systematic action, possibly in the form of efficient auditing institutions that can point out major

Table 5.1. *Indicators concerning Women in MENA and Other World Regions*

Region	Females in labor force (percentage of total labor force) (SIMA, 2001)	Female literacy rate (percentage of females aged 15 and above), 2001 (ILO, 2002)	Secondary school enrollment of females (percentage of gross enrollment) (SIMA, 1998)	Tertiary school enrollment of females (percentage of gross enrollment) (SIMA, 1998)
MENA	26.51	66.91	65.07	24.31
East Asia	42.79	87.87	77.23	26.44
South Asia	36.35	44.74	46.05	3.20
Europe and Central Asia	46.78	96.35	86.77	41.03
Latin America	35.66	85.03	73.99	28.54
Africa	43.55	58.60	31.53	3.10
G7	43.42	99.73	103.90	58.28
Western Europe	41.79	98.63	114.74	52.39

Source: World Bank Institute.

Box. 5.1. *Bureaucratic Obstacles to Entrepreneurship*

The estimated total savings of Egyptians, Jordanians, and Syrians in OECD countries has been reported to exceed the GDP of those countries (Page, 2002, p. 74). The lack of a sound investment climate in the region has primarily hampered transfers of money back to these countries. Bureaucratic hurdles and notoriously slow bureaucratic processing play a significant role.

In the region, administrative discretionary powers and inadequate regulations impede business by encouraging government interference in economic activity, prevent new and small entrepreneurs from entering the market, and are costly for international, national, and local entrepreneurs. For example, it usually takes six weeks, eleven steps, and a good 35 separate documents to set up a business in Morocco. Once these requirements are met, entrepreneurs need to obtain approval from local, central, and/or government agencies as well as utilities companies, which often exert discretionary powers.

An inadequate legal framework, insufficient reliability of the judiciary and a poor record for respect of property rights and contract enforcement have further deterred investors and entrepreneurs from starting new companies in many MENA countries.

Source: Based on World Economic Forum, 2003.

---

obstacles to change, and reforms backed by appropriate enforcement instruments can help to improve the situation. A dynamic civil society with active business associations would also help.

Trade integration might encourage reform of regulatory frameworks and diminish administrative discretionary powers. In Egypt and Jordan, provisions have been drafted for the establishment and governance of companies, following a recommendation to conform to European Community practices especially for the governance of joint stock companies, minority protection, and governance of holding companies.

State budgets are affected by the large numbers of public enterprises that have yet to be privatized. The privatization process is slow, compared for instance with eastern Europe where the transition has been more brutal. While the pace has helped maintain employment levels, it does not encourage economic efficiency and transparency.

In a number of countries, special zones have been established to attract foreign investors or facilitate trade. These initiatives are useful and help to provide new jobs, technology transfer, and so on; they are counterproductive to the extent that they prevent nationwide efforts to improve an overregulated institutional and legal framework.

A high level of corruption is one of the most dramatic consequences of bureaucratic economies. In some MENA countries, corruption is a serious problem. For instance, bribes to customs' agents often increase trade transactions by 20 percent or more (Hoekman and Messerlin, 2002). More generally, corruption is one of the most important governance issues in the MENA region (table 5.2) and needs to be addressed by all available means.

Finance

In most MENA countries, the functioning of financial markets leaves much to be desired. Banks do not operate freely and are often obliged to support public enterprises, stock exchanges are underdeveloped, and foreign exchange is controlled. Modernization would give banks more freedom and security and allow them to provide businesses with more diversified services. Countries may take excessive precautions for outstanding engagements and thus have excess liquidity. They are also reluctant to make capital assets fully convertible on more flexible foreign exchange markets because they lack hedging instruments and the qualifications needed to use them.

Table 5.2. *Good Governance Indicators*

Indicator	Low income	Lower middle income	MENA region			World average	Upper middle income	High income
			Min.	Average	Max.			
Voice and accountability	31.5	45.4	1.1	32.6	77.6	50.4	61.3	82.0
Political stability, lack of violence	27.7	42.5	4.3	49.7	95.7	50.4	63.2	86.4
Government effectiveness	26.5	42.5	3.8	53.8	88.8	50.5	62.7	87.4
Regulatory quality	28.5	42.9	1.8	49.4	82.8	50.6	65.6	83.4
Rule of law	27.0	42.6	1.8	55.7	84.1	50.5	62.7	88.6
Control of corruption	27.8	43.9	5.0	51.6	83.9	50.6	60.9	86.3

Source: World Bank Governance Indicators, 2001 (http://www.worldbank.org/wbi/governance/data.html).

The development of such financial services is an essential part of the knowledge economy and a source of jobs for a highly qualified labor force. It is also an infrastructure of key importance for economies aspiring to enter the information age. Egypt and Lebanon have made significant progress towards modernizing their financial infrastructure and may inspire others. Part of the solution lies in attracting foreign banking and insurance partners. Moreover, a modern financial system helps to improve the overall governance climate, including the establishment and enforcement of the rule of law.

New entrepreneurs require venture capital. The development of a vibrant private sector is impossible without a financial sector that is willing and able to take risks and invest freely in new firms. Microfinance support is often enough to start businesses. Existing initiatives in the MENA regions as in others should be carefully studied as ways to scale up microfinance resources. A tradition of risk sharing in Islamic banking may also lead to original ways of providing risk capital (box 5.2). Governments may wish to encourage reforms in this direction.

Trade and Economic Integration

Trade

The Muslim world played a central role in the development of the Silk Road, which might be viewed as the first instance of globalization. The inclination to trade remains very perceptible domestically, but MENA countries have unfortunately been left behind by the recent globalization process because of their poor investment climate.

Trade of goods, particularly capital goods, is an important source of knowledge transfer and acquisition. The MENA region is the region that is least integrated in world trade flows. However, tariff and nontariff barriers have been lowered in recent years, and a number of agreements with the European Union should help. Trade liberalization is not enough, however. Bureaucracy and corruption must be reduced, and transport infrastructure, telecommunications, and services, notably financial services, must be modernized. In addition to taking action at their borders, countries must take action internally (Hoekman and Messerlin, 2002).

MENA countries have little incentive to engage in intraregional trade, given that a number are largely oil producers. There are however opportunities for trade integration when economic structures are complementary (box 5.3). The gradual shift to more diversified bases of production, with finer product differentiation and design and other improvements from a knowledge-economy perspective, should encourage integration throughout the region, even for products which appear to be the same today.

Box 5.2. *Islamic Banking*

Modern Islamic banking began in Egypt in the early 1960s, and the creation of the Islamic Development Bank in 1973 gave considerable credibility and visibility to Islamic finance. Its main principle is the prohibition of *riba* (interest). Islamic banks are similar to credit unions. Savings are placed on a Mudarabah basis, which remunerates account holders by sharing the bank's annual profits rather than by earning interest.

The rationale behind the prohibition of *riba* is that profit sharing is more equitable than interest-based returns, the fluctuations of which often relate not to the contract between the bank and its depositors but to changes in monetary policy. Profit sharing has the double advantage of remunerating the depositor in relation to the bank's performance and involving the depositor much more closely with the bank and its risks. By adopting the principle that finance should be asset-based and not interest-based, Mudarabah posits that the return to the bank is justified by ownership risk for assets acquired or held, rather than by the capacity to advance funds in conventional trade financing.

Despite the competitive advantage the Arab world might gain from Islamic banking, the weight of Islamic finance in the Middle East remains modest. Since the early 1970s, efforts have been made to attract more clients by diversifying Islamic financing through various financial innovations. Despite such efforts, reluctance to engage in long-term funding has been the main obstacle to the development of Islamic banking. The size of Islamic banking, and Arab banking in general, remains modest and implies that it is difficult to leverage local capital for ambitious projects. Large-scale projects like Saudi Arabia's development of the gas sector often bypass Arab funding and are arranged through international banks.

Source: Rodney Wilson in World Economic Forum, 2003.

Box 5.3. *Trade Integration and Comparative Advantages between Lebanon and Syria*

Syria is endowed with natural resources, land, water, mineral resources, and a young and growing labor force. It has a comparative advantage in agriculture and industry, including the potential for developing agribusiness. It has invested in infrastructure but requires financial resources and substantial investments.

Sector	GDP Structures in 2000 (percentages)	
	Lebanon	Syria
Agriculture	12	24
Construction and industry	26	30
Services	66	46

Lebanon has a comparative advantage in banking and financial services and can provide the financial resources Syria requires, and Lebanese banks have been set up in Syria. This opening up to foreign financial services, international banking, and financial markets through the Lebanese banking system will create healthy competition, help introduce the required innovations in payment systems, banking, and financial services, and ultimately create closer economic integration between the two countries. Increased specialization and the signing of an association agreement with the EU would also lead to better integration into EU supply chains.

Source: Saidi, 2002.

Economic Integration

To help MENA countries, particularly those located on the Mediterranean basin, to develop and become better integrated in the world economy, the EU MEDA program was launched in Barcelona in the mid-1990s to strengthen the support provided by European countries. The program has so far not achieved the hoped-for results, and the need for greater support is now being considered.

The MEDA program is a bilateral program based on association agreements signed between the EU and 12 Mediterranean countries. Only 15 percent of the total budget is reserved for regional projects. Under the Barcelona program, the EU provides the MEDA countries US$30 per inhabitant over the next five years (compared to US$600 per inhabitant for countries that are to enter the enlarged EU in 2004). To compensate, the European Investment Bank has opened a credit facility, which may eventually become a financial institution for the area.[2] In any case, better economic integration is important for strengthening MEDA countries' links with the EU—an important market—as they face greater competition from the new EU countries that benefit from EU "structural funds."

Attitudes to Knowledge, Innovation, and Management

The production, dissemination, and application of knowledge depend primarily on a society's attitude to knowledge and its willingness to endorse it unambiguously. In fact, a piece of information only becomes knowledge once the content has been experimentally validated, codified, and reassessed. In this respect, the Arab science of the past should be a source of inspiration; if properly reactivated and supported it can become a source of rejuvenation, growth, and competitiveness for the peoples of the region.

A certain understanding of democracy is required to develop the critical minds that are a fundamental source of knowledge production and to adopt the ethics of demonstration and proof, which alone ensure scientific progress and protect against obscurantism. A society's communication practices must also support the transmission and dissemination rather than the restriction of knowledge. This is a problem in the MENA region, where information is an attribute of power and is distributed reluctantly.

The question of innovation also needs to be addressed. There is a certain ambivalence about the concept of innovation, which is rooted in a misinterpretation of religious precepts and is made worse by linguistic ambiguity. The word "innovation" can be interpreted as *bidaa*, which is frowned upon as a negation of the precepts of the Koran, or as *Ibdaa*, which is encouraged as a way to renewal and enrichment. It is obvious that those who interpret innovation as *bidaa* oppose innovation and may even rise against it.

To advance the knowledge economy, it is necessary to address the issue of the influence of traditional mindsets and behavior on the management of businesses.[3] The patriarchal model, which gives the chief a kind of absolute power, still affects relationships with authority. The search

---

2. One may wonder whether, in the longer term, an institution could be established on the pattern of the European Bank for Reconstruction and Development, which was decisive in facilitating the economic transition in a number of eastern Europe countries after the fall of the Berlin Wall. The role of such an institution would be to support in particular the emergence of a dynamic private sector by providing financial assistance and other services to critical masses of small and medium-sized firms and by connecting them to trade, training, and other types of networks vital to their expansion.

3. See the presentation by R. Benmokthar of Al-Akhawayn University at the Marseilles conference, September 2002. These elements are based on a detailed firm-level inquiry made by the *Centre de Recherche des Dirigeants d'Entreprises* and published in Mezouar, 2002.

for consensus and avoidance of conflict weigh on decision making. Because entrepreneurship is not perceived as the result of individual effort initiative is stifled; moreover, the enterprise tends to be seen as exogenous to the society.

Mindsets and worldviews do not evolve quickly. They will change gradually as modernization takes place throughout the society. Change can be accelerated by appropriate discussion and self-analysis in receptive groups. These include entrepreneurs and managers confronted with economic competition, which challenges traditional views and habits. Professional and trade associations can usefully stimulate such discussions (Mezouar, 2002). Over the long term, schools, beginning at the primary level, will also play an important role.

6

Education and Training

Knowledge is essentially transmitted through initial education and training and lifelong learning. It can be argued that an effective education and training system is the basis of progress, as it forms the human resources that produce knowledge, the labor force that employs it to produce goods and services, and the individuals who use it in their daily lives. A knowledge-based society and a knowledge economy will not develop if a large share of society cannot retrieve, select, and interpret the information that is constantly disseminated worldwide. This chapter discusses a few of the main challenges facing the countries of the region.

An educational system must provide: (i) a broad-ranging culture largely based on the traditions and the features of the society, which teaches the meaning of things in a given context; (ii) socialization in an equitable manner, on the basis of a widely accepted meritocracy, in which equal chances are granted to the greatest number; and (iii) training in accordance with labor market demand. The MENA region's educational systems are still mainly oriented toward the first two functions of education—human and social development—but neglect the third, the aptitudes needed to perform well in the job market.

With few exceptions, today's education systems have difficulty reconciling these three objectives. In industrialized societies, the difficulty is due to rapid changes in demand on labor markets, while education systems, by the nature of their basic objectives, tend to evolve relatively slowly. Europe, for example, lacks engineers and scientists, and the relations between schools and firms are the subject of recurrent debate. MENA countries face more fundamental difficulties from the cultural, social, and economic viewpoints.

Cultural and Linguistic Issues

The various elements that make up a culture should encourage a synthesis of the three functions that are the essence of education systems. In MENA countries, however, culture does not yet play this role. In fact, it often plays a "defensive" role, and to protect a sense of identity it may (in its extreme forms) glorify the past and divide society between those who embrace "modernism" and others. This situation is largely due to the fact that these countries have still only partly incorporated their culture into the material reality of society. In brief, having gained their independence and consolidated the state, they need to make their cultures less virtual.

The arabization of education offers an excellent example. It is being carried out under conditions that handicap an entire age group. Because some countries did not choose a language of reference, particularly for the scientific and technical disciplines, they now have a weak and somewhat incongruous system. Pupils do not master correct Arabic or the foreign language most widely available (French in the Maghreb and in Lebanon, English in the other MENA countries). This is largely because of: (i) a lack of harmonization in teaching Arabic; (ii) a lack of translations of reference works, especially scientific ones, into Arabic; and (iii) the poor organization of what could be a real bilingual capacity (experience overwhelmingly shows that learning a second language must take place intensively and very early). Bilingual studies dominate at the primary and secondary levels in Tunisia. In Algeria, the curriculum is entirely in Arabic except at the higher levels. In Egypt, Jordan, and Morocco, the schools in which the teaching is in Arabic are separate from those in which teaching takes place in a foreign language.

In its present form, the region's education systems seriously handicap pupils and students. They are also an important factor of social segregation. While education ought to be a basis for progress and cohesion, only the children of the elite master a foreign language and can pursue their studies in Europe and in North America.

Education and Social Expectations

A second set of issues relates to social expectations. Essentially, the education systems of MENA countries aim at mobility and social integration, not at meeting the needs of the economy. After independence, they made a considerable effort to generalize schooling. Given their spending on education, they have been ranked among the most committed to education. Schooling rates have increased considerably, and it is becoming universal in the leaders. In Tunisia, for example, which ranks first in the MENA region from this point of view, the gross schooling rate is 117 percent in primary schools (against a MENA average of 96 percent) and 55 percent in secondary schools. At the other extreme, Morocco's aim is universal schooling for those aged 6–15 years old by 2006; however, despite state spending of more than 7 percent of GDP on education, gross schooling rates are 80 percent at the primary level and only 39 percent at the secondary level.

This state of affairs draws attention to a number of issues:

- The qualification delivered by the education system is a social qualification which is often sanctioned by a general diploma. It is based on a system in which the number of years of schooling counts more than the development of specific competencies. This is typical of economies in which a large public sector delivers basic public services.
- Professional education and the validation of professional achievements are rare or non-existent.
- From the bottom to the top, the education system sanctions failure, and failure at school creates the conditions of further social exclusion. Students repeat classes at high rates at the primary and secondary levels; achievements at the end of compulsory education are not acknowledged (except in Tunisia, which established the EFEF, an examination to mark the end of basic education, in 1998); the first university cycles have high failure rates.
- There is a tendency to reproduce past learning, as teachers (who do not receive continuous training) teach what they learned as students. Classical subjects (literature, law, economics) predominate to the detriment of scientific and technological disciplines. There are too few short university courses (university training cycles tend to range from four to seven years). Educational institutions are in fact largely isolated from the rest of society, are not systematically evaluated for performance, and do not address the market and its needs.
- The private sector is underrepresented in educational institutions (except in Jordan).
- Education systems suffer because teachers' salaries decrease in real terms, resources for students are reduced in areas that need to be developed (taking into account the needs of the knowledge economy), and libraries, computers, and multimedia facilities are under-equipped.
- Education systems will face increasing difficulties in the coming years, considering the current level of education budgets, as they must improve performance in terms of universal schooling while facing demand for increasingly higher levels of training.

Training and Job Markets

A third set of issues pertains to labor market demand. Most teachers take the view that it is the economy's fault if young people are unemployed. This attitude reflects many of the shortcomings mentioned above and is justified. First, growth in most of the region's economies is insufficient to meet the growing labor supply (table 6.1). Furthermore, the region's economies are not yet sufficiently diversified and developed to offer jobs to all graduates. The share of unemployed young people is worrisome everywhere. Although sectors like ICT in particular lack the necessary workers, training/employment mismatches make it difficult to absorb the unemployed.

Egypt is a good example. In the most optimistic scenario, the supply/demand ratio for a secondary school and university graduate is 400 percent. This means that it would take 20 years for graduates of secondary schools in Egypt (2 million) to find jobs, given average annual growth of employment (10 percent). Egypt tried the German system of dual learning with little success because it lacked industrial technicians and positions were not available for half of the apprentices. In general, unemployment of young people has increased sharply (more than 30 percent in Morocco during the past 20 years, a phenomenon that has also been observed in Algeria, Tunisia, and Egypt), and it seems to be higher for graduates than for other categories.

Finally many employers, mainly in the informal sector, prefer to hire unqualified workers and relatives who accept hard working conditions and low salaries. The same phenomenon, in a less exacerbated form, is apparent in the domestic industrial sector where, in view of technological obsolescence, the need for more qualified labor is not clear. This may explain the refusal of educational institutions to be led by uncertain demand.

It is therefore understandable that there is little vocational training and especially lifelong learning. In initial education, pupils who find general education difficult are relegated to vocational training, which does not allow for access to university (except in Jordan where vocational training is very developed). This situation is difficult to correct given the large numbers of students already at the secondary level. Moreover, vocational training never feeds into continuous training, except in the modern and technological sectors of the economy and those with links to international firms.

In Tunisia and Jordan, which have opened up the most and have most diversified their productive sector, there is now a net demand for vocational training. It is reckoned that every year the Tunisian labor market requires 60,000 workers with average qualifications, while only 30,000 vocational secondary level students have such qualifications. There are three reasons for this. First, wage workers reject vocational jobs and prefer public administration and general training. Second, low-status vocational training is insufficiently promoted and lacks links to other

Table 6.1. *Need for Economic Growth and Growth of the Labor Supply, 1990–99*
(percent)

Country	Increasing need for jobs (labor growth rate) (i)	GDP growth, consistent with (i) (ii)	Observed GDP growth (iii)	Gap (iv) = (ii) – (iii)
Algeria	3.8	5.4	1.5	3.9
Egypt	2.9	4.1	4.3	—
Iran	2.3	3.3	4.0	—
Jordan	5.8	8.3	5.2	3.1
Kuwait	5.1	7.3	2.1	5.2
Morocco	2.5	3.6	2.2	1.4
Saudi Arabia	3.1	4.4	2.1	2.3
Syria	4.3	6.1	5.9	0.2
Tunisia	2.9	4.1	1.8	2.3

Source: Keller and Nabli, 2002.

forms of training. Third, the vocational training system is very centralized and rigid and is poorly adapted to actual market conditions. These remarks are true of all countries of the region and become increasingly pertinent as schooling rates increase and the industrial structure diversifies.

In recent years, however, the MENA region's need for vocational training is acknowledged everywhere, and efforts have been made to improve the situation. Tunisia's MANFORME program aims to raise the quality of vocational training in closer cooperation with the productive sector and with greater private sector participation (box 6.1). In Jordan, vocational training is integrated in the classical secondary system (three types of training are proposed: short, medium, long). The combination of classical education and apprenticeship leads to a degree (General Secondary Certificate in Vocational Education) which allows entry to university. In Egypt, diversification of technical and vocational education is under way and some courses (via the Technical Secondary School) also lead to postsecondary technical training.

These efforts go in the right direction but are still too isolated, fragmented, and rigid. Their social and economic value is uncertain because they are poorly evaluated. They also run up against the social primacy given to years of education over learning and the obvious lack of communication between the worlds of education and of firms, to the detriment of meeting economic objectives. The problem is aggravated by a perception that the training is too limited to improve prospects and is not, on the whole, motivating given the need for additional effort and no clear rewards.

These remarks take on greater importance for lifelong learning. The knowledge-based society can only grow if every individual is keen, throughout his or her lifetime, to acquire new knowledge and to update knowledge acquired earlier. In a freer and more open system, this makes it possible to adapt to new labor market conditions and to acquire the necessary mobility to "cover" the risks of participation in the globalized world. This is also what makes it possible to participate

---

Box 6.1. *The Reform of Professional Training in Tunisia*

Tunisia's reform of professional training (MANFORME) is a national strategy adopted in 1995 for a period of 10 years by the Ministry of Employment and Vocational Training with the support of the World Bank, the European Union, and other international organizations. In a context of decreasing tariff barriers and opening of markets, notably the establishment of a free trade zone with the European Union as of 2008, the reform of professional training demonstrates Tunisia's efforts to prepare for international competition.

The current reform of the education and training system began in 1991 with the establishment of compulsory basic education. It was followed in 1993 by a Law on Professional Training, which defined a new paradigm in which education and training are adapted to economic demand and the structure of Tunisia's labor force converges with Europe's. Europe's pyramid of worker qualifications is typically 20 percent low-skilled, 60 percent medium-skilled, 20 percent high-skilled. The distribution in Tunisia is 60 percent, 30 percent and 10 percent, respectively.

To better understand companies' needs, professional structures are fostering a new relationship between the administration and economic actors. Continuous training in a firm is replacing alternative training. A decentralized approach facilitates cooperation between training centers and firms.

Certain issues need to be addressed to achieve the goals of this paradigm. There should be more flexibility in terms of advancement, which remains based on seniority, and in the regulations governing teaching and administrative staff. Moreover directors of training centers should have more budget autonomy in order to maintain the centers' innovative capacity. Finally, partnerships with firms could be strengthened through greater involvement of professional employers' organizations and Chambers of Commerce and Industry.

Source: Bensaid, 2002.

in progress, whether technical or organizational. The experience of the industrial countries, where this is an obligation, shows that lifelong learning depends on the individual's will to continue to learn. This requires raising awareness, professional and social incentives, a diversified and easily accessible network of supply structures that eliminate the separation of schools and firms, and professional and social recognition of the efforts made.

MENA countries are still far from offering an effective framework for lifelong learning. Aside from the present limits of continuous vocational training, lifelong learning focuses at best on the narrow fringe of "intellectual" workers who follow an unorganized path of self-training.

An Education Strategy within a Lifelong Learning Perspective

A lifelong learning strategy should mobilize both the public and the private sectors. They should define in common the basis for building an efficient lifelong learning system (qualifications, accreditation, and so on). Related services can then be usefully provided by the private sector, so long as a certain level of equity is maintained.

Principles

For lifelong learning, it is first necessary to have a foundation of basic skills in order to profit from more technical and professional training, even through the use of "unfettered" educators such as the media, especially television. The region is still considerably behind in terms of both "stocks" and "flows." Many people in MENA countries are illiterate in terms of reading and writing skills and even more have not mastered the new "literacy criteria," that is, the capacity to understand and synthesize texts and computer skills. The illiteracy rate is 50 percent in Morocco and close to 40 percent in Algeria and to 30 percent in Tunisia, and illiteracy rates among women are particularly high. Despite huge financial efforts, these rates diminish very slowly because the struggle against illiteracy is not sufficiently widespread and because system failures still produce illiterates.

The solution is twofold: (i) the struggle against illiteracy must intensify, as some countries (for example Brazil) have done, so that every pupil at the end of compulsory education has at least a reasonable set of basic skills (which they can subsequently improve); (ii) the technical or vocational skills related to the qualifications required for the degree should be enhanced. All experiments have shown that it is illusory to expect to generalize lifelong learning by immersion or even by using significant training credits (vouchers) for long-term institutional programs that are sanctioned by diplomas. These are generally reserved for the managerial staff of large firms and cannot be generalized for many reasons: the costs are extremely high, immersion isolates staff from the professional milieu, there is a lack of agreement between the professional and the educational worlds, training institutions are concentrated in big cities, and so on.

For lifelong learning, maximum use should be made of ICT, to better disseminate teaching content and lower the cost of accreditation (which is very high everywhere) and, concomitantly, to develop an educational software industry which MENA countries other than Egypt are lacking. Finally, continuous training must affect salaries and promotions, especially in the passage to managerial levels. This implies that the professional branches and unions must be associated in the strategy to develop lifelong learning.

Codifying Knowledge and Encouraging Citizens to Acquire It throughout Life

If the objective is for the greater part of society to have access to knowledge, clear and easily verifiable targets must be set to ensure the commitment of educational institutions at the various levels at which they play their role. Later, individuals must have incentives to respond to a range of educational offers (educational institutions, training centers at different levels, self-learning with the help of ICT) because of a desire for personal growth and social recognition. It may be that the knowledge economy will emerge as a result of social demand.

First, there must be commitment at the highest levels so that all pupils have, at the end of compulsory education, a basic set of skills (speak, read, and write the mother tongue; basic understanding of a foreign language; arithmetic and computer skills) that will allow them to continue their education and training. This requires a clear definition of requirements (which a cohort should be able to master); these should not be the norm but a minimal reference. The means to guarantee that all students can master these skills should be available (through special readjustment and second chance courses). Mechanisms, notably using ICT, should be created and supplied so that adults can also gradually obtain and validate such skills.

Second, knowledge (general and technical/professional) should be codified. Codification implies segmenting knowledge into modules (or credit hours) and is indispensable for mobility between countries (through the accumulation of credits variously obtained) and for lifelong learning (if the codification is properly done, it gives access to clearly specified knowledge without lengthy stays in educational institutions).

For general knowledge, an effort could be made to define a common body of knowledge for MENA countries to be used as a reference for certain key stages (and act, among other things, as a humanistic reference) such as the end of compulsory or secondary schooling. If mathematics, sciences, and computerization do not require a search for specific solutions to achieve a certain standard, such a search is necessary for history, citizenship, moral values, and, most of all, the Arabic language. The content would not necessarily focus on the region's concerns but could be part of "the mosaic formed by all the human cultures" (Serres, 2002). This is a gigantic task, but it is precisely what the Muslim world achieved in its golden age.

For technical and vocational knowledge, codification should be carried out on the basis of notions such as competence, qualification, vocational skills, and the correspondence among them, so as to promote cultural assets that can be combined by the individual, in accordance with his/her experience, and that are socially and professionally recognized. This implies serious reflection about knowledge itself in order to: define professional profiles and related skills; modularize general education (to favor the earning of qualifications in different stages); set correspondence grids in place for skills, qualifications, degrees, and professions; develop instruments at different levels for the validation or accreditation of technical and vocational skills; set up a system to issue degrees for professional skills acquired and for a composite of more limited skills; constitute mixed working groups of schools, firms, vocational branches, chambers of commerce and unions, to define interlinked grids, and reexamine technical and vocational paths and possibilities for general education.

Increasing the Efficacy of Educational and Training Institutions

The efficacy of educational and training institutions can be appreciated from an internal or external perspective. How these institutions function internally depends on their capacity to fulfill their traditional missions. For the most part, they are public institutions, and evaluation of their performance should be reinforced, generalized, and made public. The evaluation should be carried out by an external body and focus upon simple and unambiguous criteria. Particular attention must be paid to:

- Deficiencies at the primary levels in some countries, which necessitate the decentralization of primary schools. World experience suggests that such decentralization can be efficient when local governments exercise their power appropriately.

- The insufficient number of scientific and technical subjects, a structural aspect that can probably be corrected by introducing elements of science and technology (especially computer literacy) into the arts, the humanities, and the social sciences, and by introducing management courses for science students.
- Failure in the first university cycle, which must be addressed by transfers to "colleges," with teachers who may not have the same status as teaching researchers.
- The need for better evaluation of performance in vocational training centers and of what students learn in these centers (this requires verifying the value-added aspect of training). Existing accreditation of these centers should not be relied on.
- Generalizing operations to ensure quality standards in educational institutions.

External evaluation will develop if a market-based approach is introduced into education and training. This approach, recently adopted by some countries, can make additional resources available to educational and training systems. It is a way to develop new activities focused on innovation and training with firms and based on demand. This entails:

- Greater penetration of the private sector into education and training in competition with the public sector;
- Greater autonomy of establishments so that they can keep and freely use the resources resulting from their research activities, consulting and training, have greater control over their budgets, increase their registration fees, and allocate these new resources to the teaching staff, so that they are better paid and able to reduce their outside activities.
- Making known the results of professional integration in the different institutions and allowing for freer choice of schools and universities.

7

Innovation and Research

Poorly Developed Innovation Systems

Promoting a technologically creative and innovative economy represents a considerable challenge. One can hardly argue that MENA countries have, at present, true innovation systems founded on extensive interaction among research, production, and training. Some elements are there, fragmented policies have been set into place, but clear structures, with competitive assets, are lacking. Three categories of countries can be distinguished.[1]

The first includes Algeria and Egypt. These countries have made serious efforts to integrate science and technology in development policies. However, they have not done so on the basis of a coherent plan and have not engaged in a process of decentralization that would encourage firms to make regular use of scientific and technical progress. In the 1970s and 1980s, Algeria developed extensive scientific, engineering, and technical training and established many turnkey public enterprises. It therefore had human capital that was better trained to face industrial realities than other countries of the region. However, this potential was little exploited because of these firms' monopolistic position and the rigidities of the regulatory environment in which they operated. Delays in restructuring and privatizing these public enterprises have hindered making full use of these capacities, and except in hydrocarbons, the national innovation system is presently stagnating. Egypt has emphasized higher education and produces highly qualified and specialized personnel. The government launched large-scale basic and applied research programs, and Egypt now has many R&D institutions. They have been remarkably successful in areas such as agriculture, inorganic chemistry, and pharmaceuticals. However, the innovation system remains fragmented and very much centered on these institutions.

A second group of countries, including Morocco, Tunisia, Jordan, and Kuwait, has adopted more open and demand-oriented processes and has attracted foreign establishments and developed subcontracting activities. They have developed science and technology in specialized areas (free zones, technological poles) but have not made this part of a comprehensive development policy. The creation of engineering schools (polytechnics) (especially in Tunisia) was a positive step, but the excessive importance given to training in the humanities and social sciences has limited the necessary production of scientific workers.

The industrial base of countries in the third category, which includes Syria and Yemen, is insufficient to develop an innovation system. These countries also lack the necessary infrastructures.

R&D Institutions

The overall R&D effort in the MENA region is only 0.3 percent of GDP (the world average is 1.4 percent). The number of researchers per total population, measured in full time equivalent (FTE), is less than one-tenth the number in OECD countries. Moreover, only a small share work in the business sector.[2] Nevertheless, the region has many R&D institutions, especially in universities. The

---

1. This section relies heavily on Djeflat, 2002.
2. According to 1996 data, the number of FTE in the MENA region was 0.3 per 100,000 inhabitants and 4.6 per 100,000 for OECD countries (19,000 FTE in the Arab world in 1996). Only 2.1 percent were in the business sector, while 66.3 percent were in government institutes and 31.6 percent were in universities.

number of universities in the region has increased significantly from 10 in 1950 to 177 in 1996, and the number of R&D institutions has grown by about eight a year since the beginning of the 1960s. Today, there are 322 R&D institutions (against 26 in 1960), but they tend to be isolated from the economic and social world. Many bottlenecks still exist.

Problems range from the low priority accorded to R&D, to the lack of a strategy for developing a coherent scientific and technical framework, to the insufficient development of private research, to the institutional rigidities that paralyze private initiative, such as excessive centralization bureaucratic control. Budgetary ups and downs (especially as a result of the impact of monetary depreciation) also create constraints and slow equipment purchases. Staff regulations limit mobility and incentives to perform. The statute governing university teacher-researchers is a central problem, as it leads to very modest educational performances in the first university cycle and poor research output (university personnel spend less than 10 percent of their time on research).

Some countries (notably Egypt, Jordan, and Kuwait) have made a special effort to create bodies to coordinate public research in universities and industry. However, interaction and linkages are still lacking or are incomplete where they exist. Research is carried out for the most part in government and university laboratories. Applied research focuses on the medical sciences and agriculture, and only a small part is devoted to industrial technologies and the new areas of information, molecular biology, and genetic research.

Innovation Climate in Industry

Although the MENA region offers a significant number of success stories (box 7.1), receptiveness to innovation and industry's innovative capability are generally low (Djeflat, 2002).

First, the labor force's technical skills are poor. Two-thirds of the working age population do not have the requisite training. Only a small number of engineers and holders of science diploma work in the business sector, and many are unemployed. Technical supervision is insufficient. Imported equipment is much less productive than it should be. In Egypt, for example, the productivity of imported equipment is estimated to be only 50 percent of what it is in the advanced countries of origin, because of the labor force's poor qualifications. Moreover there is resistance to change and innovation both from workers on the shop floor who are afraid of losing their jobs and from top management which is poorly informed or very conservative.

A lack of competition, reinforced by protectionist policies and/or a large state industry, does not stimulate innovation and the upgrading of output. New and small firms encounter bureaucratic and regulatory obstacles and difficulties for accessing finance for innovation and expansion. When financial support effectively exists, only a small minority benefits because of a lack of information or the complexity of the procedures involved.

The quality and breadth of technological services are insufficient. Professional services are imported by MENA countries in amounts that are equal to or larger than imports of capital goods, although most could be provided locally. Infrastructure for technical assistance, quality control, and so on are lacking in most countries, although they are in great demand, as shown by the rapid adoption of ISO standards by all sectors, including small-scale ones. In Algeria, for example, about 70 percent of small firms are reported to have adopted ISO standards.

Links with foreign firms are mainly established through the purchase of equipment and services by public sector enterprises. When there are links between foreign firms and other domestic firms, the impact in terms of innovation and upgrading of technological competence is limited because the domestic firms are generally low on the value chain and employ cheap, unqualified labor to carry out subcontracting activities.

Box 7.1. MENA Success Stories in Innovation

Development of R&D Departments in Tunisian Industry

A recent survey of the Tunisian chemical industry presents an encouraging picture. Out of the 31 firms with R&D departments in the chemistry sector, 60 percent (19 firms) have modified their products. Two-thirds of these believe that the presence of an R&D department has led to higher turnover by improving processes (70 percent) or by improving products, hence the realization of new products (54 percent). For the most part, these services are relatively recent (they have existed for about 10 years). Managers admit that the R&D units are essentially created to maintain the capacity to compete. Other studies in various sectors also show that the success of R&D units contributes to acquiring new equipment, modifying technological processes, improving or introducing products, and improving marketing. Internal interest in R&D has also grown, and this strategic element has been taken over by management. Directly available internal and external sources of information have been used (symposiums, seminars, clients), while the chambers of commerce and the national centers set up by the government (CETTEX, CETIME, CNCC) were judged only marginally useful. All the enterprises concerned have increased their employment rates.

Innovation through Public Institutions: Kuwait's KISR, Egypt's ASRT

The Kuwait Institute for Scientific Research (KISR) and the Egyptian Academy of Scientific Research and Technology (ASRT) are among the most successful public research institutions in the MENA region. Over the past two decades, KISR has achieved significant results in water resources, textiles, and energy savings. Kuwait is considered one of the first countries in the world to have mastered the technique of water desalination. KISR was established with help from Japan in 1972 and has survived because it has very clear and specific objectives which are regularly evaluated. Active and sustained international cooperation helps to master state-of-the-art technologies and adapt them to local needs. The Egyptian ASRT has also obtained significant results through scientific and technological cooperation programs, using local raw materials, especially for paving roads, and developing computer and food production technologies. These results were made possible by high motivation, appropriate incentive systems, and competent implementation of R&D projects.

Small Firms in Algeria

The remarkable development of small firms in Algeria was made possible by the recent adoption of a more liberal policy (especially for imported equipment and goods). About one-third of these firms specialize in construction and public works, 16 percent in trade, and the rest are divided between transport and communication, services to persons, and agriculture. Despite modest access to banking credit, these enterprises have made important investments (between 1994 and 2000, investment projects were multiplied by 17) and generated more than a million jobs. Despite the difficulty of obtaining credit, these (mostly micro) enterprises have improved the quality of their products, invented new ones, and replaced equipment. This shows the importance of liberalizing the regulatory framework to generate innovation.

Innovations by Public Enterprises

Large public enterprises can also be a source of innovation, especially when they are responsible for the production and dissemination of a natural resource (such as the OCP in Morocco or Sonatrach in Algeria). But other, less prestigious firms can also progress if they are encouraged to do so by the authorities and if they are appropriately managed. SAIDAL, a pharmaceutical company in Algeria, provides a convincing example. This firm, created in 1998, efficiently developed generic medicines and captured 40 percent of the Algerian market. Great attention was paid to quality control (ISO 9000) and R&D. ISO recognized the quality of the firm's products, and it was able to export almost immediately (to Iraq, Italy, Senegal, and South Africa, among others). As a result, turnover was higher during the first semester of 2002 than in 2001.

Source: Djeflat, 2002.

Scientific and Technological Performance

MENA countries' industrial competitiveness remains low, although Jordan, Tunisia, and the United Arab Emirates have export growth rates above the world average in some manufacturing sectors (textiles, electronics) (see the country profiles in World Economic Forum, 2003). The share of capital goods in MENA exports is very low (except in Tunisia where it reaches 35 percent). The share of technology and skill-intensive products in total exports is also low (25 percent or less).

Deferring the adoption of innovations means that capacity is out of date. This threatens the competitiveness of important sectors. In Tunisia, textiles account for almost half of the workers employed in industry, but only 20 percent of the spinning machines are fully automatic. The remaining 80 percent use obsolete mechanical technology. In Algeria, various estimates indicate that electronic equipment and agricultural machines last about 25 years; in Egypt, the problem is the same in several sectors.

The standard indicator, patent applications, is still very low. Over 400 patents are filed annually by Algeria, 97 percent of which by foreign firms (12 patents were filed by local firms). The numbers are increasing in Tunisia (141 in 1995, 257 in 2000, and 178 in 2001), in Morocco (325 in 1996), and stand at 410 a year in Egypt (but only 10 percent are of local origin). However, this is still far from the standard in emerging countries like Turkey (722 patents filed annually at the end of the 1990s).

Scientific publications are at a very low level. The ratio of scientific publications in refereed journals is 20 per million inhabitants in the Maghreb (26 per million for the Arab world as a whole) as compared to 42 in Brazil and 144 in Korea. Furthermore, only 5 percent of the articles are about engineering and technology, and these are not published in reputable international journals.

Establishing a Coherent Innovation Policy

In more than one MENA country, governments have made significant efforts to strengthen their science and technology (S&T) policies, increase resources devoted to S&T, establish priorities, create focal points in the form of centers of excellence, and develop technology parks (for an overview, see Djeflat, 2002). These efforts are meritorious, but more comprehensive action is required to improve the innovation climate. Contrary to a well-established view, research does not precede innovation. In fact, the path is often the reverse: technical advances lead to new lines of research. This is particularly the case in developing countries. The priority therefore is to establish efficient mechanisms that support innovators, their initiatives, and the products they design and wish to diffuse in the economy.

Developing Support to Innovators

In addition to measures to improve the general investment climate and to remove regulatory and bureaucratic obstacles to innovation, a first priority is an efficient infrastructure that provides basic technical services in the form of quality control, certification, and standards—in particular, internationally recognized standards such as ISO 9000. This can facilitate the design and marketing of new technologies and ensure their diffusion, including to export markets (see the example of Algeria in box 7.1). Tunisia has launched noteworthy initiatives in this area, but in most MENA countries this infrastructure is extremely weak. Certain established public research institutes could be officially "certified" to provide the required quality control and assurance.

In addition, technical information networks and assistance centers are needed throughout the country (or at least in the principal areas of activity) to help innovators to design, develop, and test their products. Incubators that offer entrepreneurs common services at a reasonable price are also needed. They can be a source of marketing, business, and financial advice and may facilitate access to venture capital. They are generally located in, or close to, university campuses or in science and technology parks whose role is to facilitate the formation of local critical masses of new enterprises (an essential feature of dynamic innovation climates). They are generally more efficient if they are run by businesses or associations, although appropriate government incentives can facilitate their development, as in Egypt.

To defend their interests, new and small firms also require support from professional associations. Such support is particularly useful in the bureaucratic and corrupt environments that often exist in MENA economies and favor large, established enterprises linked to power structures.

Finally, and this is crucial, efficient protection of property rights must be ensured. The quality of the property rights regime is problematic in certain MENA countries, and it is necessary in particular to reduce the production of illegal imitations (counterfeit products) (Maskus and Penubarti, 1995). The access of developing countries in general, and MENA countries in particular, to patented technologies is an important issue as well, as the cost is prohibitive for low-income and medium-income countries (World Bank, 2001, chapter 7). This may lead to a certain lack of discipline, as demonstrated by countries like Brazil or South Africa for drug-related patents.

Tapping into Foreign Technology and Knowledge

Foreign direct investment is the primary means of taking advantage of foreign technology. Experience shows, however, that the firms that benefit most from FDI are those that supply parts and components to foreign subsidiaries involved in the assembly and production of finished goods (as recently demonstrated for Indonesia by Gertler and Blalock, 2001). This is because these suppliers receive strict guidelines regarding quality, standards, and delivery conditions. They also benefit from (or are obliged to acquire) sophisticated equipment and machinery. Another benefit of FDI is the impact on the skills of employees of foreign subsidiaries and of their suppliers. However, as noted above, most MENA countries still receive very little FDI. This again shows the importance of a systemic approach to improving the overall investment climate. Appropriate regulations and R&D structures are needed so that countries can benefit from technology transfer from foreign firms and increase local innovative capabilities. However, while FDI has increased (albeit slowly) over the years, the number of patents filed by residents has diminished as a percentage of total patents filed.

A second way to acquire the necessary knowledge and technology is to seek them out, particularly in advanced countries, and make appropriate arrangements for international cooperation, joint ventures, licensing, and so on. Countries such as Korea and Chile have systematically developed networks and organizations for this purpose. Some MENA countries are also occasionally involved in such practices, notably in the oil sector. Agriculture and water management present similar opportunities. Efficient public research institutions seem to be an appropriate vehicle for tapping into global knowledge (see the Kuwaiti and Egyptian examples in box 7.1), provided that there are then adequate mechanisms for transfer to the private sector.

Human resources constitute the third, and probably most efficient, way to tap into foreign knowledge. Dubai, for example, which has an aggressive development strategy, has tried to attract a highly trained and competent foreign workforce to help to develop new fields such as software or the media. There is also an important diaspora of highly qualified expatriates who are currently employed by firms, universities, and laboratories in advanced countries. According to some estimates, there are more than 1 million highly qualified Arab expatriates in OECD countries (UNDP, 2002, p. 71). MENA countries, following the example of countries such as China, may take measures to attract them home. Experience shows, however, that as long as the general environment for working, innovating, and conducting business is problematic, and more generally until they can enjoy a standard of living comparable to that in the advanced economies, expatriates do not return.

Adapting R&D and R&D structures

R&D structures are in poor condition in most MENA countries. They suffer from underfunding, poor links to industry, and a highly constraining and bureaucratic environment. Various measures need to be envisaged.

Business R&D should be stimulated by appropriate financial support, notably to small firms, by subsidies to innovation-oriented projects, by recruitment of researchers, and so on. It is also important to encourage collaborative research between the business sector and universities and public research institutes, for example through incentives for joint programs and subsidies for contract research commissioned by firms.

There should be clear incentives in favor of research and its applications. These include providing aid for filing patents in the U.S. and European patent offices, enhancing the value accorded to applied research when evaluating researchers, making it possible for researchers to start innovative enterprises, facilitating the mobility of researchers to the private sector, creating fiscal incentives relating to the products of applied research (such as lower taxes on royalties for computer software), and improving the conditions for collaboration with foreign partners.

The national R&D development strategies should draw on recognized strengths and needs. Some sectors deserve a strong research base in order to pursue their development or to further exploit their comparative advantage. Oil-related sectors are of course a case in point, but MENA countries have also acquired some comparative advantage in water desalination, specific pharmaceuticals, and agriculture products (such as cotton).

Public research bodies should have greater autonomy at least with respect to their use of own resources. The research system needs to be "deregulated" to allow it more easily to take initiatives, and this should be accompanied by clear and independent mechanisms for evaluating them. Criteria and mechanisms should be different for programs or centers involved in applied or technical research and for those involved in basic research. The former should be judged according to criteria of relevance and the latter according to criteria of excellence.

The conditions for international research collaboration should be improved and benefit from appropriate bilateral and multilateral support. International cooperation can play a central role by helping to supply scientific journals and manuals in cooperation with European and American editors, by developing operations and programs that help to spread know-how, and by providing computer equipment and software for purposes of distance learning.

Promoting an Innovative Culture

Promoting an innovative culture requires efforts that affect the entire social fabric. A scientific and technical culture needs to be developed, beginning at primary school levels but also encompassing the population at large (see a discussion of the Portuguese experience in chapter 9). Creativity should be promoted in university curricula and evaluated. Greater emphasis and value should be placed on technical jobs than on administrative jobs in the administration and in firms. Finally, the overall climate for becoming an entrepreneur and setting up and expanding a business should be better. This begins by removing undue regulatory and bureaucratic obstacles.

Conclusion

For all of these types of measures, there is an abundance of well-established good practice in the international community (especially in OECD countries; see OECD, 2000). It should not be difficult to adapt them to the MENA context if the existing administrative frameworks become sufficiently flexible. It would however be useful to consider the establishment of strong innovation promotion agencies with adequate resources, an effective staff, and sufficient authority to intervene wherever and whenever necessary in the innovation process. OECD countries offer various models, which may help MENA policymakers meet various objectives and needs. For instance, France's ANVAR has efficiently supported the creation and growth of innovative firms (notably on a regional basis). Finland's TEKES has very effectively promoted research collaboration and innovative programs involving university and public laboratories on the one hand and the business sector on the other. The Fundación Chile has been a remarkable instrument for the growth of competitive traditional sectors.

8

Telecommunications and the Information Infrastructure

Telecommunications and Related Policies

The telecommunications infrastructure plays an essential role in the knowledge-based society. It is also indispensable in the struggle against the digital divide and for breaking down the isolation of rural areas. Furthermore, the telecommunications infrastructure is essential in the MENA countries that are experiencing conflict, as it can replace transport. This is an area in which most countries of the region still lag behind. It appears necessary to: (i) reinforce investment in order to improve the population's access to these services; (ii) improve service quality, which suffers from many problems (cuts, losses, high occurrence of faults per line, and so on); (iii) markedly lower the cost of access and use; and (iv) build the human resources and skills to support the infrastructure and related service offerings.

From the end of the 1990s, almost all MENA countries speeded up telecommunications reform, and their regulations are now comparable to those of other emerging countries. However, equipment levels and usage rates are generally low for traditional fixed-line services. The mobile sector grew rapidly in countries that introduced competition in that area. In fact mobile penetration has caught up with that of fixed lines. However, the private sector's success with mobile telephony has hindered penetration of the Internet in MENA countries. Internet access is related to the availability and cost of fixed phone lines, since most technologies for accessing the Internet are based on fixed voice telephony circuits. Wireless technologies are more expensive, and the entry barrier is therefore higher. Internet access is as low in the MENA region as in Africa, despite much higher GDP per capita. MENA Internet users represent only 1 percent of the world user base (Network Users Access Internet Survey, September 2002). This has a significant negative impact on trade, education, research, and access to innovation networks.

Table 8.1 shows, for example the high cost of local phone calls, which are only a part of accessing the Internet (see table 8.2). One needs a PC, a phone line, an Internet account and, of course, available power.

Policy Trends

Policy trends in most MENA countries address restructuring of the telecommunications sector, privatization of the incumbent national operator, and development of mobile networks.

Restructuring is under way almost everywhere in the region. It mainly involves separating management of postal services from that of telecommunications, separating political decision making from economic regulations through the creation of independent regulatory agencies, and clarifying and specifying the role of the incumbent operator with respect to new actors.

Since 2000, two new regulatory authorities have been created in Algeria. Lebanon is also setting up a regulatory authority. These countries have thus followed reforms undertaken in Morocco, Tunisia, Jordan, and Egypt. Countries such as Syria, Iran, West Bank Gaza, and Yemen have still a long way to go in this area.

After a period of reorganization and regulation, many countries plan to gradually achieve full liberalization. This is the case for Turkey, which wishes to enter the European Union, but also of Morocco, Jordan, Egypt, and Algeria.

Table 8.1. *Cost of Communications, Line Density by Country, 2000*

Country	Telephone mainlines				Mobile international		
	Number of telephone mainlines (per 1,000 pop.)	Waiting list (per 1,000 pop.)	Waiting time (years)	Cost of local call (US$ for 3 min.)	Number of mobile phones (per 1,000 pop.)	Outgoing traffic (minutes per sub-scriber)	Cost of call to U.S. (US$ for 3 min.)
Algeria	57	646.0	5.4	0.01	3	86	4.70
Egypt	86	1,300.0	1.9	0.01	21	34	3.33
Iran	149	1,203.5	1.2	0.01	15	24	7.65
Iraq	29	—	—	—	0	29	—
Israel	482	—	0.3	0.05	702	324	3.30
Jordan	93	29.7	0.3	0.02	58	275	—
Kuwait	244	0.0	0.0	0.00	249	340	5.41
Lebanon	195	—	—	0.07	212	124	4.45
Libya	108	80.0	1.2	—	7	78	—
Morocco	50	5.0	0.1	0.07	83	172	4.50
Oman	89	—	0.5	0.07	65	518	—
Saudi Arabia	137	927.4	2.6	0.01	64	324	5.20
Syria	103	3,025.8	> 10.0	0.02	2	101	20.04
Tunisia	90	83.7	0.9	0.02	6	165	—
United Arab Emirates	391	0.3	0.0	0.00	548	1,102	3.51
West Bank Gaza	—	—	0.7	0.04	—	—	—
Yemen	19	159.5	3.8	0.01	2	105	4.45

— Not available.
Source: World Bank, Global Information and Communication Technologies Department.

The partial privatization of incumbent operators has also gained momentum. It began in Jordan (where it has been completed), and Morocco followed in 2000. The partial privatization of Morocco Telecom, like that of Turk Telecom, resumed in 2002/03. In Algeria and Lebanon, laws governing the sector's reform explicitly provide for opening up the incumbent's capital to a strategic foreign operator. A law passed in Tunisia in January 2001 simplified the conditions for foreign participation, although this has yet to happen. Most countries of the region began quite late to pour investments into telecommunications, but they have already reached significant rates of digitization. The capital assets of the incumbent were often opened to private foreign firms because of the need to modernize the infrastructure to improve penetration of fixed telephony lines. It was observed that investing in fixed lines was costly and might affect the balance of the incumbent operator's account, as it did in Jordan, which has led the way in terms of universal access. In Jordan, a project of the incumbent, with government support, equipped more than 350 rural villages and created more than 200,000 new telephone lines. The project was considered a total success (installation of switching capacity, increase in transmission equipment, doubling of the number of subscribers, mobilization of the capital market by issuing bonds), but it has led to a 13.5 percent increase in current expenses and only a 5 percent increase in annual income. The gap partly resulted from the modification of the system of cross-subsidies. It is thought that this might be compensated for by the development of mobile telephony.

International experience has shown that a good regulatory agency is crucial for forcing potential operators to fulfill their universal access obligations by including such clauses in their license requirements. To enter new markets, investors require good license design and appropriate bid design (see the example of Morocco in box 8.1). Several MENA countries have suffered from the

Table 8.2. *MENA Region: Teledensity and Internet Penetration*

Country	Per million population 2001	Fixed lines per 100 inhabitants 2000	Mobile lines per 100 inhabitants 2001	Internet users per 10,000 inhabitants 2000	Internet hosts per 10,000 inhabitants 2001
Maghreb					
Algeria	31.88	5.70	0.32	16.19	0.01
Libya	5.27	10.79	0.57	17.84	0.11
Morocco	30.77	5.02	1.45	70.54	0.21
Tunisia	9.74	9.00	0.32	104.32	0.11
Mashreq					
Egypt	69.82	8.63	0.45	70.89	0.83
Jordan	5.19	9.29	1.42	190.87	1.83
Lebanon	3.64	19.49	2.12	858.00	18.12
Syria	16.83	10.35	0.10	18.53	0.00
West Bank Gaza	3.50	—	—	—	—
Gulf States					
Bahrain	6.63	24.97	4.39	584.19	1.82
Djibouti	0.46	1.52	0.01	21.94	0.02
Iran	66.36	14.90	0.22	39.27	0.12
Iraq	23.33	2.94	—	—	0.00
Kuwait	2.06	24.39	4.47	783.53	16.56
Oman	2.65	8.88	0.73	354.59	2.44
Qatar	0.78	26.76	1.92	501.29	0.00
Saudi Arabia	22.93	13.72	1.01	92.56	4.11
United Arab Emirates	2.42	39.14	7.25	2,820.46	132.00
Yemen	18.21	1.88	0.06	8.17	0.04

— Not available.

Sources: International Telecommunications Union, European Mobile Communication Group, Internet Software Consortium.

poor quality of the bidding process, license design, and contractual arrangements with private sector operators under Build–Operate–Transfer (BOT) arrangements (as illustrated by the cellular sector disputes between the two mobile operators and the government in Lebanon).

Finally, in all countries of the region, growth of telecommunications and better access to telephone services have been mainly due to the development of mobile telephony by private investors. In Jordan, Algeria, and Syria, new private operators obtained licenses to operate mobile telephone networks (BOT). New global system for communications (GSM) licenses are to be granted in Algeria toward the end of 2003 and in Lebanon in 2003. The Tunisian Ministry of Communication Technologies, after having decided not to grant a second license (the first private project was launched in March 2001), is again studying the conditions for the installation and operation of a second GSM network.

The Internet

Cost is still the main obstacle to Internet access. In addition to the cost of communications and of purchasing a computer, subscribing to a service provider is the main impediment today. The limited number of sites in Arabic and the poor quality of the infrastructure and technologies available add to the problem.

Box 8.1. *Morocco: Successful Regulation Enables Growth of Telecommunications*

The regulatory framework is crucially important to investors in emerging markets. A successful privatization process requires accountability, credibility, transparency, and efficient regulation. Morocco is one of the few MENA countries in which the auction for the second national operator (for the second mobile GSM operator) was acclaimed as a success, both in the bidding process and the end result. The 1997 Telecom Act, implemented in 1999, made the ARNT the telecommunications regulatory watchdog and established its autonomy from the government. Competition was introduced in 2000 for a second national GSM operator with the following parallel processes: corporatization of the incumbent operator (Ittisalat Al Maghreb), separation of the policymaking function from the regulatory function: ANRT, licensing regime for new operators, obligation to publish the reference interconnection offer for operators, and regulated universal service tariffs.

The license terms were based on the tender documents and on the value-added services offered by the many bidders. The selection of bidders was based on the following criteria, all given a specific grade:

- License fee: 60 points for best offer,
- Tariff structure: 15 points according to reference terms,
- Coverage and network quality: 20 points (grade according to supplement compared to minimum described in tender document),
- Offer coherence: 5 points; business plan, employment training, technical knowledge, and knowledge transfer.

The government ultimately collected US$1.1 billion in license fees, and the winning bidder included comprehensive coverage and tariff reduction commitments. The success of the granting of the second GSM license owes much to the transparent and open conditions under which bidding took place. Many of the bidders indicated that Morocco's clear regulatory structure and its political stability reduced investment risks and helped them increase the value of their bids.

The drawback to the mobile phenomenon in Morocco today (25 percent of the population has a cellular phone) is the decline in fixed phone access, which has repercussions on Internet access and use. Despite a 27 percent (annual) growth in mobile subscriptions in 2001, Morocco's fixed telecommunications sector suffered a 12 percent drop in customers. This will affect Internet use and penetration. With still high entry costs (cost of local calls, personal computer acquisition costs, Internet account costs) and the lack of content that is relevant to most of the population, growth of the Internet in Morocco continues to suffer, in part from the success of the mobile market.

However, several commissions or national pilot groups have been set up in recent years to establish strategies for developing the information society. This is especially true of Syria, Tunisia, Jordan, and Turkey; Saudi Arabia and Kuwait are ahead of the others. Strategic orientations and specific programs have been formulated to encourage widespread use of the Internet in public administrations, education and research, and even in medicine.

Commitment can also be measured by the opening of markets to Internet access providers. Today, all MENA countries except Syria have at least partially liberalized this market, and the number of access providers is rapidly increasing. However, there still are some constraints.

First, a newcomer still has to obtain licenses and authorizations in all countries (especially in Jordan, Algeria, and Egypt). Most Internet service providers (except in Egypt, Lebanon, and Morocco) have to go through a government-controlled International Gateway. Also, the Internet service provider is not allowed to build its own network infrastructure and hence pays heavy fees to rent the infrastructure from the incumbent operator or, when available, from the second national operator. Finally, connectivity costs are too high for businesses to use the Internet as an ordinary business tool.

In addition, some governments are wary of offering national information online and try not to lose full control of content, even on existing national sites.

Providing Access for the Many Modern Forms of Communication

Although the most important telecommunications reforms are undoubtedly under way, the authorities must take further measures in two directions. They need to invest heavily to improve the quantity and quality of infrastructures. This is partly the state's responsibility, at least in the short term. Also, they need to pursue the liberalization process, so as to obtain a tangible reduction in both access and usage costs.

The main services available are local services, national long distance intercity services, long distance international services, leasing of fixed lines for dedicated Internet access and data communications at a reasonable cost, and mobile telephony.

The cost of telecommunication services has come down markedly, and there is no reason to maintain a monopoly, given the capacity of optic fibers. The development of the knowledge economy has been slowed because prices are high, not because of costs but because national operators set high prices for long distance services (particularly international services) to subsidize local services (they cross-subsidize). Obtaining authorization to provide rented services for distance telecommunications is complicated, time-consuming, and hard to obtain, and once obtained, the quality of service is low. Connection between mobile and fixed telephones has been slowed by contentious interconnection agreements. New licenses aimed at reinforcing competition did not provide the possibility of access to the Internet.

The objective is therefore clear. It is necessary to improve possibilities for telephone communication (such as new fixed lines), to lower the cost of access to telephone services and to the Internet (and develop inclusive billing), to generalize the use of leased lines to businesses to facilitate telecommunications-based work (distance subcontracting), and to develop e-commerce. In a benchmarking exercise for five MENA countries it was found that the cost of a local 64 KB leased line is generally 12 times higher than the OECD average.

The need to increase competition through appropriate regulation implies a number of things:

- Renouncing the financial spoils of a national operator that monopolizes international communications by granting other licenses and allowing other license owners access to international services in order to lower prices;
- Decreasing cross-subsidies by promoting competition;
- Allowing possessors of a mobile license to be connected to a fixed network, which will introduce competition in long-distance national lines;
- Setting up an independent and credible regulatory agency responsible for granting licenses, fixing prices (ceiling price, for example), and arbitrating conflicts between operators (in particular to develop interconnections when the national operator intervenes);
- Facilitating the availability of leased lines to allow for distance working (today this is done by a national operator that charges high costs for long waiting hours), which must be the responsibility of the regulatory authority;
- Opening up the fixed-line market for data providers to increase the appetite for Internet connectivity and broadband access and lower the prices of existing leased lines from the incumbent operator.

The MENA telecommunications market is far from saturated, and the data segment offers great potential for all actors concerned: corporate, education, government, and households. However this potential is being bridled by monopolies on fixed-line services and international access.

Building Human Capacity for Telecommunications

Given the constraints mentioned above, and with the exception of the United Arab Emirates, MENA countries today do not attract the best and brightest telecommunication engineers and information technology experts to set up large businesses and expand in local markets. Except for governments and multinationals, local markets are too small. For governments modernizing existing systems (budget, human resources, procurement, decentralization, taxation, and so on) offers promising but daunting prospects. As for multinationals, most use their own privately owned networks and bypass incumbent operators for reasons of capacity and service quality. This leaves local small businesses and households, whose entry to the information society is slowed by high entry costs. A small business in the MENA region pays on average five times more a month on a eight-hour narrowband connection to the Internet (based on a voice telephony line) than an American small business with unlimited Internet access at a speed 20 times higher (and an annual GDP per capita at least 15 times higher than in the MENA region).

These factors push bright telecommunication experts to emigrate, and in fact a large proportion of the few experts being formed in the region do leave. Post-September 11 problems such as country reputation or difficulties in obtaining visas are reshaping the picture, but the brain drain is still strong from MENA to Europe and to the United Arab Emirates, where Dubai's Internet and Media City (see chapter 9) is a pole of attraction for new graduates.

A final issue to recall regarding the development of telecommunications in the MENA region and the related building of knowledge societies is the cultural mindset inherited from decades of colonialism and still repressive regimes in several countries. This mindset has instilled a fear of communicating and a lack of information sharing. The separation of genders continues as women remain secluded. Innovative programs for using telecommunications and information technology to provide knowledge for all need to be put in place and supported at the grass roots level.

9

Visions and Strategies

The MENA region faces many challenges and is reaching a critical point in its history, as forcefully demonstrated by the recently issued *Arab Human Development Report* (UNDP, 2002) and unanimously agreed by participants at the Marseilles conference. It was also admitted that it is not so much a question of resources, but a matter of new approaches in the conduct of societies and economies. As Jean-Louis Sarbib, Vice President of the World Bank for the MENA region, pointed out at the overall conclusion of the conference, the region needs an enlightened and strong leadership, able to develop and carry out new, appealing visions in which the very sense of freedom is central in order to unleash the region's creative and entrepreneurial potential. Some inspiration can be found in other parts of the world as well as in the region itself, as country presentations at the conference showed (Sarbib, 2002).

Selected Country Experiences

Malaysia

Malaysia, a country with a strong sense of identity, offers a convincing and original experience of continuous economic growth (John, 2002). After decades of laissez-faire economic policies and export-driven growth, Malaysia designed a development strategy based on economic and rural development and socioeconomic equity that has prevailed for the last 30 years. It has achieved average growth of 7 percent over the 1960–2000 period. It successfully migrated from the agricultural age to the information age in one generation (1970–2000). Absolute poverty has decreased from 49.3 percent to 7.3 percent in 30 years.

From 1971 to 1990 Malaysia implemented a new economic policy with growth with equity as its central concept. From 1991 to 2000, it had a balanced national development policy. From 2001 to 2010 it aims to build a resilient and competitive knowledge economy. In the late 1990s, when it was severely affected by the Asian financial crisis, Malaysia decided to leap into the knowledge age.

It developed an ambitious ICT program that was designed to drive growth and bridge the digital gap. The program included the building of a "Multimedia Super Corridor," the establishment of many ICT centers throughout the country (40 information technology community centers are accessible to 1.2 million people), and investments in education and human resources. The overall plan, a part of the National Information Technology Framework, is to develop simultaneously human resources, infrastructure, and content. New forms of governance are envisaged based on the gradual integration of the public, private, and community sectors. The proposed vision clearly reaffirms the foundations of a pluralistic, democratic Islamic nation that adopts universal values. The overall vision and strategy are mainly developed and carried out by the National Information Technology Council of Malaysia, located at the level of the prime minister and operating across ministries. The choice of the knowledge economy was the result of both a vision and a crisis which was perceived as an opportunity for change.

However, the efforts made to move toward a knowledge economy need to be sustained. The important investments made so far have yet to prove their value as a way of allowing Malaysia to continue to perform economically within the broad social balance that it has sought.

Korea

Korea offers another example of a knowledge-based development strategy to overcome problems experienced during the 1998 Asian financial and foreign exchange crisis (Kang, 2002). In the wake of the crisis, Korea became particularly receptive to this new development concept, and the Korean president launched a national knowledge-based strategy. In April 2000, led by the Ministry of Finance, and supported by a World Bank-OECD report, the strategy evolved into a three-year action plan for five main areas: information infrastructure, human resources, knowledge-based industry, science and technology, and elimination of the digital divide. To implement the action plan, five working groups were formed involving 19 ministries and 17 research institutes. Every quarter, each ministry submits a self-monitoring report and mid-term results are sent to the National Advisory Council, which coordinates Korea's overall economic policy.

Korea's main business newspaper, *Maeil*, played a key role in raising awareness both in the policymaking community and society by launching, prior to the 1998 crisis, a knowledge economy conference with international experts, which it has since repeated every year. Orchestrated by the newspaper, strong media campaigns were launched to promote and reward knowledge workers, stimulate mind change in firms, administrations, schools, and even the army. Reports were commissioned to international consulting firms to help diagnose the strengths and weaknesses of the economy and its different segments as they relate to the knowledge economy.

New activities related to high technology and related services spread quickly and broadly, reducing the unemployment rate from 7 percent in 1998 to less than 4 percent in 2002. The contribution of the information technology (IT) industry to GDP is estimated to have increased from less than 15 percent in 1996 to more than 50 percent in 2000. Growth of GDP is expected to reach 6 percent for 2002. Foreign reserves have been reconstituted to record levels (more than US$100 billion in 2001 from less than US$10 billion in 1997). These successes were made possible by the country's heavy investments in education, R&D, and technological infrastructure since the early 1980s.

Portugal

After a smooth transition from an authoritarian to a democratic regime in the mid-1970s, Portugal has gradually built a competitive economy. Benefiting from the active support of the European Community, it has massively invested in education, R&D, and other knowledge-related areas. As a result of its economic modernization and performance, Portugal saw its GDP per inhabitant more than treble between 1981 and 2000. In current purchasing power parities, Portugal's GDP per inhabitant was three-quarters of the OECD average in 2000 against only half of the average in 1980. The dimension of this shift in economic focus is also reflected in the dramatic change in the structure of the labor force between 1974 and 1998. Currently more than half of the labor force works in the tertiary sector as opposed to one-third in the early 1970s (Gago, 2002).

Central to this economic transformation has been an active policy to develop the population's scientific and technical culture through many different initiatives, most importantly at the school level. In 1996 the *Ciencia Viva* ("Science Alive") program was launched in three areas: science education in schools, science and technology awareness of the general public, and interactive science centers. It has received 5 percent of the Ministry of Science and Technology's budget (€13 million in 2001). The initiative concerning science in schools supports practical activities in secondary education and involved, up to the end of 2000, more than 3,000 schools, 7,000 teachers, and 500,000 students. The program has been strengthened by twinning schools and scientific institutions to promote joint activities and exchanges of knowledge. At present, 37 schools and 20 scientific institutions are involved.

As in Malaysia, such initiatives have been crucial for interesting the population in a new model of development based on knowledge, information, and innovation. Moreover, at the European level, Portugal has gained considerable respect and visibility by positioning itself as the main architect of the e-Europe plan adopted by the EU Heads of State in 2001 as the foundation of European countries' competitiveness in the future.

To a certain extent, experiences that are taking shape in the MENA region present similarities with those sketched out above, although they are adapted to the peculiarities of the region's sociocultural and industrial context. The following section offers three examples of how visions are being formulated and implemented at the national, city-state, and institutional levels.

MENA Countries

Jordan: A Comprehensive Strategy

Jordan has formulated and initiated at the highest levels a strategy with a nationwide perspective. Its strategy for a competitive knowledge economy is built on three elements: a national strategy with the support of King Abdullah, a comprehensive program targeting every pillar of the knowledge economy, and strong involvement of and cooperation between the public and private sectors.[1]

High-level commissions were created to advise and monitor progress, and several regulatory and monitoring authorities were established: the Economic Consultative Council (ECC) monitors the implementation of government strategies; the Jordan Authority for Enterprise Development (JAED) advises the minister on developing enterprises and improving the investment climate; and the Information Technology Association of Jordan (INTAJ) represents and promotes domestic software and IT services industry in the global market.

Initiatives addressing the four basic knowledge economy pillars were undertaken to improve the overall investment climate, to develop technological-industrial zones, to reform education, and to establish a large network of IT community centers. Change is becoming visible in the form of new enterprises and new jobs. The most significant initiatives include:

- The Jordan IT Community Centers (JITTC), created to bridge the digital divide in local communities, develop the use of ICT for social development, enhance the role of rural communities in economic development, and promote sustainable development. Currently 20 centers have been established and are operational.
- The Economic Opportunities for Jordanian Youth (INJAZ), funded by the U.S. Agency for International Development and supported by a national media campaign. INJAZ has tackled two important challenges. First, the combination of a very young population and university curricula that are ill-adapted to the demands of the global market has contributed to an increase in structural unemployment. Second, inequality between schools in the capital city and in the rest of the country has created a two-speed education system. INJAZ works with the private sector to boost Jordan's competitiveness by directing its labor force into training and educational programs adapted to market demand as it also works to reduce social and economic marginalization.

---

1. This partnership was essential for the success of policy initiatives and initiatives benefiting from active support by international donors (notably from the United States and the United Kingdom).

- The Transformation of the Red Sea Port of Aqaba into a Special Economic Zone (ASEZ). The objective of the project, which began in mid-2000, was to bolster the national economy, cut unemployment rates, and attract US$6 billion in foreign investment by establishing a liberal, low-tax, duty-free, multisector development area. The city of Aqaba became a low-tax hub with special customs and legal systems and an export-processing zone with integrated multisector growth clusters. The ASEZ Authority functions as a statutory institution empowered with regulatory, administrative, fiscal, and economic responsibilities. Financially independent, it is governed by six ministerial-level commissioners responsible for regulatory or operational activity.
- The REACH initiative (Regulatory Framework, Estate Infrastructure, Advancement Program, Capital, Human Resource Development) was launched in October 1999 to develop an export-oriented IT services sector to allow Jordan to become a regional leader and internationally recognized exporter of IT products and services.

Dubai: An Emirate Moving toward the Knowledge Economy

Dubai shows that it is possible to formulate a vision to be reached in several stages. Its transition from an economy dependent on natural resources to a full-fledged knowledge economy originated with a vision shaped and developed by Dubai's chief executive officer. The vision embraced the knowledge economy as the means of changing Dubai's growth model.

As in most Middle Eastern economies, a considerable decline in income from energy exports (table 9.1), rapid population increase (5.8 percent a year), and limited FDI were impeding Dubai's economic growth and development.

Dubai's development was envisioned as the creation of services, products, applications, and, most important, employment in the areas of IT, the Internet, data processing and telephony. Foreign competencies were attracted through appropriate immigration laws, and evaluations by international auditing firms benchmark the city against other attractive places of world. The sheik who leads the city is the key mover.

As its main economic objective, Dubai projected a 30 percent increase in GDP within 10 years, of which 25 percent in the knowledge economy and 70 percent in the overall services sector. These achievements were to be facilitated by the establishment of an open market economy in the form of a free trade zone.

Market objectives involved transforming Dubai into a "node in global networks," a world hub to act as a regional leader and a bridge between Europe and Asia. Finally, social objectives entailed shaping a new class of knowledge workers, so that young and dynamic entrepreneurs and business leaders could contribute to Dubai's integration into the global community without loss of identity.

Dubai's knowledge economy strategy encompassed three phases. Horizon 1 focused on Dubai's assets in trade, logistics, transportation, and tourism to capitalize on "what Dubai does best."

Table 9.1. *Share of Oil Sector in Dubai's GDP, Selected Years, 1995–2005*
(percentage of GDP)

Year	Share of oil sector in Dubai's GDP
1995	18.0
2000	10.4
2005 (expected)	6.4

Source: Figures communicated by Dr. Omar Bin Sulaiman at the Marseilles conference.

This enabled Dubai to diversify production and to propel itself toward a knowledge economy. Horizon 2 revolved around the application of core competencies to new areas such as technology, financial services, media, telecommunications, and IT and was structured around technology, e-commerce, and a media free zone (TECOM) consisting of three separate business entities: Dubai Internet City, Media City, and the Knowledge Village. These entities were crucial to the success of Dubai's strategy since they were concrete knowledge economy applications in the form of visible projects. Finally, Horizon 3 built upon the revenues leveraged in Horizon 1 and 2, which were invested to project and develop future competencies. Dubai is currently looking into new developments in R&D, education, and emerging sectors such as pharmaceuticals, biotechnology, nanotechnology, and wireless.

Morocco's Al-Akhawayn University in Ifrane: Strategy Building at the Institutional Level

Morocco's Al-Akhawayn University in Ifrane illustrates an initiative taken by a progressive university to advance the country's knowledge economy agenda (Benmokthar, 2002). After undertaking a thorough analysis of the current situation, the strengths and weaknesses of the economy and society, it is beginning to work with local authorities and other partners to bring change, through a back-and-forth process between experiment and conceptual framework. The experience is expected to have a demonstration effect on broader communities, gradually creating self-confidence and removing inhibiting factors.

The approach was developed by the university's Institute of Economic Analysis and Prospective Studies and is based on a data (benchmarking) analysis of the economic situation in Morocco at both macroeconomic and microeconomic levels. It includes an original sociocultural analysis of factors of inertia in Morocco. The analysis provides a roadmap for a move toward a knowledge economy.

Topics of crucial importance for change include governance, local development, training and potential leaders, R&D, and strategic alliances at the local and global levels. In these areas, key actors are identified and mechanisms are being put in place to carry out concrete projects with those actors, with a view to removing inhibition factors and facilitating the replication of successful experiments, once they have been properly evaluated, to central and local administrations, enterprises, professional associations, universities, and NGOs.

Some Guiding Principles

The foregoing examples provide some guidance on the key factors in successful visions.

A National, Holistic Vision Formulated at the Highest Level

A first principle is to have a vision formulated at the country's highest levels. In Jordan and Dubai a holistic vision helped make entry into the knowledge economy a national priority and addressed social openness, a liberalized regulatory framework, the primacy of innovation and education/training, the development of lifelong training, and the continuous and self-sustained development of ICT. The vision should be ambitious and appealing and have credible and realistic objectives. Concrete action is then needed. Visible projects, such as Dubai's Internet City or the Aqaba Special Economic Zone, are essential to show the presence of new activities, new firms, and new jobs. It is however the improvement of the legal and institutional framework, which can unleash the country's entrepreneurial capabilities, that matters most. This can ensure the sense of freedom so essential in the MENA context.

If strong leadership is required to act as a driver and lever for change, it is also essential for the design and implementation of strategies and their components to be inclusive. In societies where a broad and deep consensus is crucial, it is necessary to proceed with in-depth consultation for important reforms and to work closely with civil society for their implementation. (Policy areas where such inclusive approaches and community involvement are typically needed include the development of basic education and literacy programs). In a similar vein, national strategies should be complemented, preceded or followed, by others formulated at city or institutional level. They serve as examples that help to build confidence in the national vision.

Efficient Implementation Agencies

A strategy for ensuring that a holistic vision pervades society must be laid down and followed up at the interministerial level. This may be possible, for example, through an agency with the authority to arbitrate in regulatory and budgetary matters, set up to support the prime minister, with representatives of the ministries concerned (education, research, professional training, telecommunications, labor, industry), as well as representatives of the professional world and educational and training institutions. Such an agency would: (i) become acquainted with all the regulatory and nonregulatory ministerial projects that are closely or distantly related to the knowledge economy; (ii) identify the levers of change, launch experimental projects, and spread good practices; (iii) put in place a systematic evaluation of existing public organizations and favor the emergence of new and private ones; and (iv) stimulate networks of institutions at the national and local levels in the different sectors concerned with the knowledge economy.

It is necessary to emphasize the financial constraints under which such strategies have to be articulated and implemented. Budgetary commitments are already high and additional financial needs that affect: (i) the budget for improving the science and technology infrastructure; (ii) the balance between the different education cycles and their budgetary allocations (with various priorities, including the fight against illiteracy and the need to maintain investment in higher education); and (iii) the provision of equipment in ICT infrastructures, but also in libraries, and computer equipment. It is not reasonable to think that the knowledge economy can be developed simply on the basis of a topdown procedure that mobilizes greater budgetary outlays. In the absence of changes elsewhere, it is likely in any case that an increased budgetary effort might not be made. Instead, existing structures should be made more efficient, and decentralized (market and nonmarket) mechanisms, which would naturally drive society to target the knowledge economy, should be developed.

Emulating Success Stories

It is necessary to publicize what works and to make successes known in administrations, firms, and associations. The most dynamic actors need to launch new efforts. Here, cities and municipal bodies can play a determining role through demonstration effects, as their populations perceive the concrete outcomes of change and take ownership of related processes. At the conference in Marseilles, Beirut, Marrakech, and Casablanca presented promising plans in this regard. Malaysia and Portugal have recognized the importance of initiatives that gain their population's belief in this new socioeconomic model of development. Underlying all of these approaches, in fact, is the idea of a positive transformation of mindsets so that the notion of change and progress gradually permeates underlying mentalities and the whole social fabric.

This is happening at the level of the MENA region as a whole. Certain countries act as pioneers by launching relatively audacious reforms in the economy, education, or telecommunications or by creating new sites (such as Internet cities). These initiatives are then emulated in the rest of the region.

Regional and International Cooperation

It would also be appropriate to articulate and develop knowledge economy visions and strategies for the MENA region as a whole. A number of institutions have achieved concrete Arab-Islamic cooperation in terms of trade, finance, and other areas (see UNDP, 2002, for a critical analysis of these institutions). Most of these initiatives however have not borne the expected fruit. The promotion of a common vision with concrete programs that mobilize the appropriate institutions can lead to new and more productive "pan-Arabism." More generally, reinforcing all forms of cooperation within the region can encourage change. In education, there are jointly developed and coordinated programs (Arab League Education, Science, and Culture Organization) and initiatives such as the Kuwait-based Arab Open University (Maqusi, 2002). Cooperation should also take place in other domains. In science for instance, and as shown by bibliometric indicators, there is much less scientific collaboration between MENA countries than between individual MENA countries and countries in the rest of the world. Trade among MENA countries is also strikingly low. Yet the presence of a language common to several hundred million people living in contiguous countries could constitute a tremendous advantage and act as a powerful factor of economic integration.

Several projects of common interest for the region as a whole were articulated at the Marseilles conference. One is the building of large telecommunication infrastructures to help connect countries via large bandwidth carriers and thereby facilitate integration in the region and the broader world. Another is the constitution of some form of joint study and monitoring institutions to support the development of innovation and research policies. An example might be the Science Policy Research Unit (SPRU) based in the University of Sussex in the United Kingdom. A third is the development of training programs for young leaders by the region's government and private sectors.

The international community could offer a great deal if its support is appropriately oriented. MENA countries are less interested in financial support (lending) than in policy knowledge and help in institutional engineering. Capacity building does not happen through direct transfer of knowledge from donors to receivers, as innumerable disappointments have made abundantly clear. A true learning process is based on several principles: ownership of initiatives by local actors (civil society, not the government alone); the systematic implication of local consultants in the implementation of new measures (not the occasional sending of international consultants); and the fact that failures should be seen as learning opportunities.

Decentralized cooperation between the two sides of the Mediterranean has strong momentum. It takes myriad forms depending on the actors involved: chambers of commerce promoting information exchanges for small businesses (box 9.1), networks of engineering and business schools, intercity cooperation for urban and management issues, and so on. In the long run, it is these decentralized initiatives, operating at the grassroots level, that will "make the difference."

Box 9.1. *SMExchange*

Supported by the World Bank, SMExchange is an example of decentralized cooperation initiatives. A knowledge management and information sharing program, SMExchange was designed to build capacity for intermediary organizations (IO) and support small and medium-sized enterprises (SMEs) in countries south of the Mediterranean aiming to join a European Free Trade zone by 2010. It has three dimensions:

- To make stakeholders in the southern Mediterranean aware of the strategic importance of strong IOs to support SMEs in their development.
- To transfer best practices in capacity building of SMEs from IOs in Europe to IOs in the Mediterranean.
- To organize the demand for and supply of capacity building programs for IOs within a market place.

The SMExchange program involves an SMExchange officer, a host IO in the south, and a portal of portals. It will begin with an investment of €2 million over three years involving four countries to be launched by the Chamber of Commerce and Industry of Marseille-Provence and Morocco (FCCISM), Algeria (CACI), Tunisia (CCIT), and Egypt (FEDCOC).

10

Conclusion

This book began by evoking the extraordinary dynamism of early Arab and Muslim civilization, which had as its foundation the ability to absorb, enrich, and disseminate the knowledge that was developing at that time. The two beacons of creativity and tolerance were Cordoba and Baghdad.

This dynamism needs to be rediscovered to overcome the challenges faced by the MENA region today. Efficient use of knowledge and innovation throughout economies and societies is the only way to restore high economic growth and create the 40 million jobs needed over the coming decade.

New development strategies must be developed and carried out, with new and better investments in education, expanded and cheaper telecommunications, and a more innovative climate. As a prerequisite, economies and societies need to be opened up by improving governance conditions and more generally favoring freedom in all its forms.

Appealing visions could be proposed to the population by enlightened and dynamic leadership. It goes without saying that the region's development would benefit greatly from an appeasement of the conflicts that it has endured for some time. Unresolved conflicts constitute fixation points which drain mental energy and divert it from the reforms that are necessary to economic and social progress. They give conservative forces a pretext for maintaining the status quo (Yacoubian, 2002).

Concrete projects and reforms make change visible and build trust. Some countries in the region have shown the way. Emulation and cooperation within the region as well as the support of the international community can significantly accelerate progress.

Today the key questions are less "what is to be done" than "how to do it." It is in this perspective that work should develop, as the World Bank Vice President for the MENA region pointed out in concluding the forum (Sarbib, 2002):

"We are going to meet in 2003 and 2004. We have a benchmark. We know where we were in September 2002. It would be very useful for those of you who would like to get something going, either on the country basis or the regional basis, to work first of all within your own country, but also to work with the partners who have just expressed their willingness to follow your leadership, and to try and commit ourselves to the challenge that next year will not let people speak or only talk about the 'what' is to be done. We will focus instead on the 'how to', and ask people to speak about how they are addressing these issues and what are the lessons they have drawn from it."

Appendix: Benchmarking MENA Countries' Readiness for the Knowledge Economy

1) Benchmarking Methodology of the *Institut de la Méditerranée*
2) Detailed Profiles for the MENA Region and Countries

Methodology Used by the *Institut de la Méditerranée* to Rank MENA Countries in Terms of the Knowledge Economy

To compare the performances of the different economies in terms of the knowledge economy, a multicriteria method is used, based on the concept of "overranking" with sorting by the "electre-tri" method of the University of Paris-Dauphine's Laboratoire d'Analyse Mathématique et Statistique Appliquée à la Décision.

The method consists of assigning each country of the sample to predefined and hierarchically organized categories or classes. Class 5 contains the best-performing countries in terms of the chosen indicators, Class 4 contains countries with weaker performances than Class 5, but higher than Class 3, and so on. A country's assignment to a category is arrived at by comparing its performances to predefined and hierarchically organized profiles on the basis of the outranking principle; that is, for a set of indicators, country A outranks profile 1 when, in "a sufficient majority" of cases (defined by the cutoff level), A's performance is at least as good as that of profile 1. When country A outranks profile 1, it is assigned to Class 5. If not, it is compared to profile 2 and is assigned to Class 4 if it outranks profile 2. If not it is compared to profile 3, and so on. When all the profiles outrank the country, the country is assigned to the last class.

To obtain robust results, there should be no ambiguity in the outranking relationship. Given the problem of imprecision in performance evaluation, it is important to determine for a given indicator when one value is better than another. Is a literacy rate of 99.0% better than one of 98.0%? To overcome the problem, the possibility offered by the method to refer to pseudo-criteria is used: the definition of an indifference threshold (Sa) and a preference threshold (Sb).

For a given criterion, the indifference threshold defines a minimum value below which a difference between the country's and the profile's performance does not matter. Starting from the previous example, if Sa is set at 1%, the two literacy rates (99.0% and 98.0%) are considered equivalent. If Sa is set at 0.5%, a literacy rate of 99.0% is considered better than one of 98.0%. However, a rate of 98.5% represents the same performance as 99.0%. For a given criterion, the preference threshold defines the value of an undeniably better performance. Sb is generally greater than Sa. In the above example, with a value of 1.0% for Sa and 2.0% for Sb, a literacy rate of 98.0% represents the same performance as 99.0%, but a rate of 97.0% (99.0%-Sb) represents without ambiguity a weaker performance than the other two rates.

Between Sa and Sb, there is an area where the outranking relation is not complete but where the situation is not the same. Referring to the notion of probability and starting with a profile value of 99.0%, an indifference threshold of 1.0% and a preference threshold of 2.0%, in comparison with a rate of 99.0%:

- A rate of 98.0% (equal to or greater than 99.0%-Sa) has a probability of 100.0% to be "at least as good as" the norm;

- A rate of 97.5% (between 99.0%-*Sa* and 99.0%-*Sb*) has a probability of 50.0% to be "at least as good as" the norm;
- A rate of 97.0% (lower than 99.0%-*Sb*) has a probability of 0.0% to be "at least as good as" the norm.

According to the norm and the thresholds, an outranking degree is computed for each country relative to each profile, which ranges from 0 to 1 and can be interpreted as the degree of reality of the assertion "this country outranks this profile." When for a given profile this degree is greater than a fixed value (the cutoff level of 70.0% in the present case), the country outranks the profile.

As this brief summary of the method indicates, the way in which the values of the profiles and the thresholds are determined is crucial for the reliability of the results. To obtain the most objective results, systematic statistical methods are used for each criterion. To compute the profile values (four profiles to divide the sample into five classes), the 80th centile of the distribution is taken for profile 1, the 60th centile for profile 2, the 40th centile for profile 3, the 20th centile for profile 4. In other words, for 100 countries, if a given country outranks the first profile for a given criterion, this means that the country is among the first 20 performers in the world (here in the KAM14 sample). On the contrary, when the country is outranked by the last profile (profile 4), 80.0% of the countries have better results than it.

The centiles of the distribution are also the basis for the calculation of thresholds. Thus, the threshold *Sa* is equal to 5% of the distribution; for the threshold *Sb*, the value chosen is equal to 7.5% of the distribution.

Figure A.1 shows that a country with a literacy rate of 84.7% is outranked (for this criterion) by the value of the profile 1 (99.0%) and the value of profile 2 (97.6%), taking into account the level of the two thresholds. This country outranks profile 4. It thus falls in Class 3 or Class 4. It is not at the top of Class 3 as the difference between the value and the profile exceeds the indifference threshold (profile 3 value: 87.4%; *Sa* value for profile 3: 2.5%; *Sb* value for profile 3: 2.95%; and 84.7% < 87.4%-*Sa*); but it is in Class 3 as its value is greater than the profile value minus the preference threshold (84.7% ≥ 87.4%-*Sb*). The point falls into the middle of Class 3 according to its outranking degree against profile 3 (linear function of *Sa*, *Sb* and the cutoff level).

Detailed Profiles for MENA Region and Countries

The attached profiles include for the entire MENA region and 11 MENA countries:

- A spider chart presenting the overall knowledge economy readiness for the years 1995 and 2001 and the 31 selected variables, and comparative spider charts for the G7, Latin America and East Asia for 2001;
- The relative position in the class breakdown figure for the years 2001 and 1995;
- Spider charts presenting the relative positions for each of the four pillars and related variables in 2001;
- A table summarizing the trends experienced for each of the 31 selected variables over the period 1995–2001.

Important note for the following tables

The evolution of the relative position takes into account the national and the international trends. Thus, a "+" indicates an improvement in the country, but also a greater improvement than the other countries of the class; a "-" indicates that the national trend has been less positive than that of the other countries, without necessarily being negative.

Figure A.1. Method of Classification: Example with the Criterion of Literacy

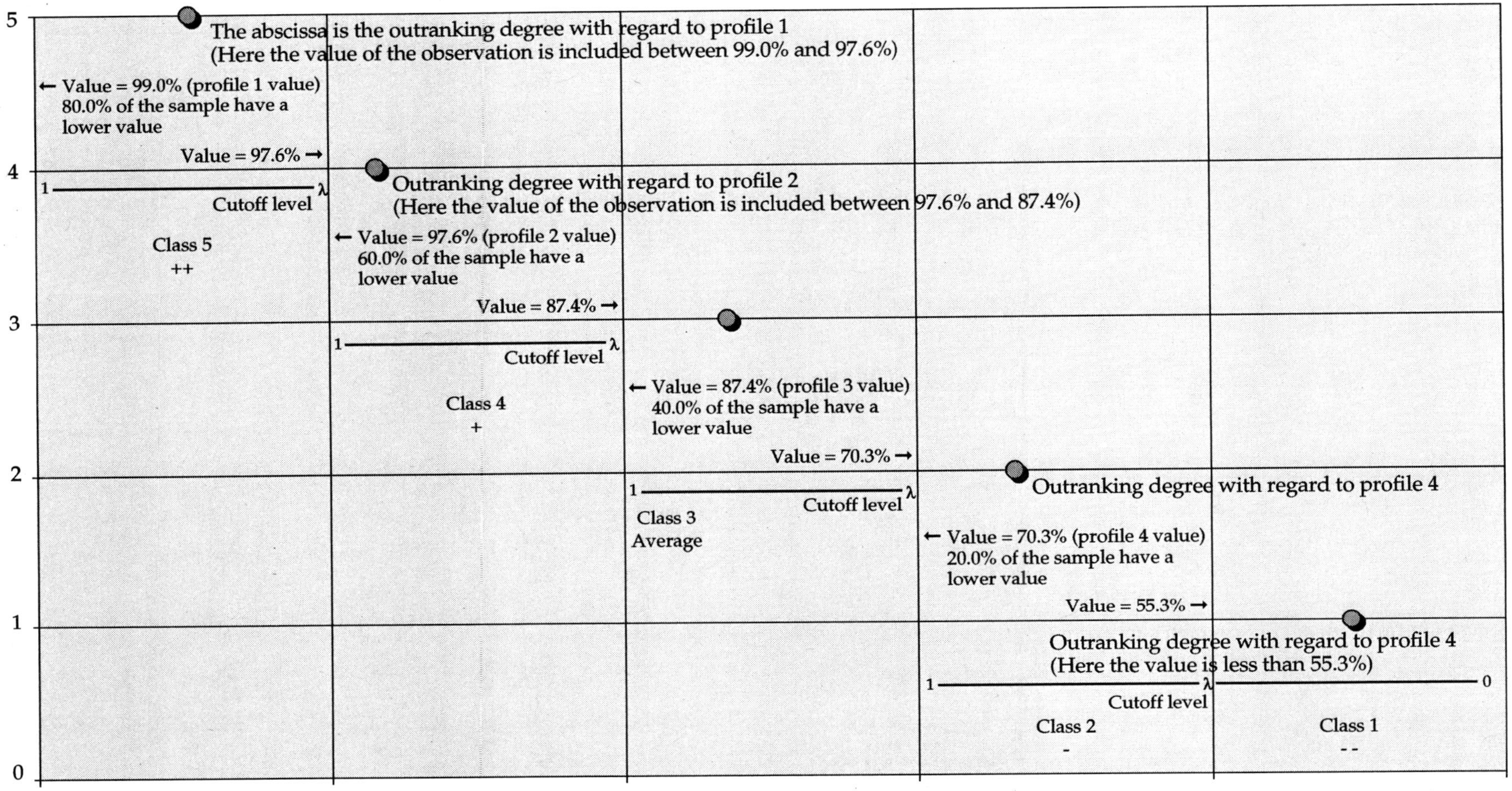

Figure A.2. Knowledge Economy Readiness Assessments for the MENA Region and for Selected MENA Countries

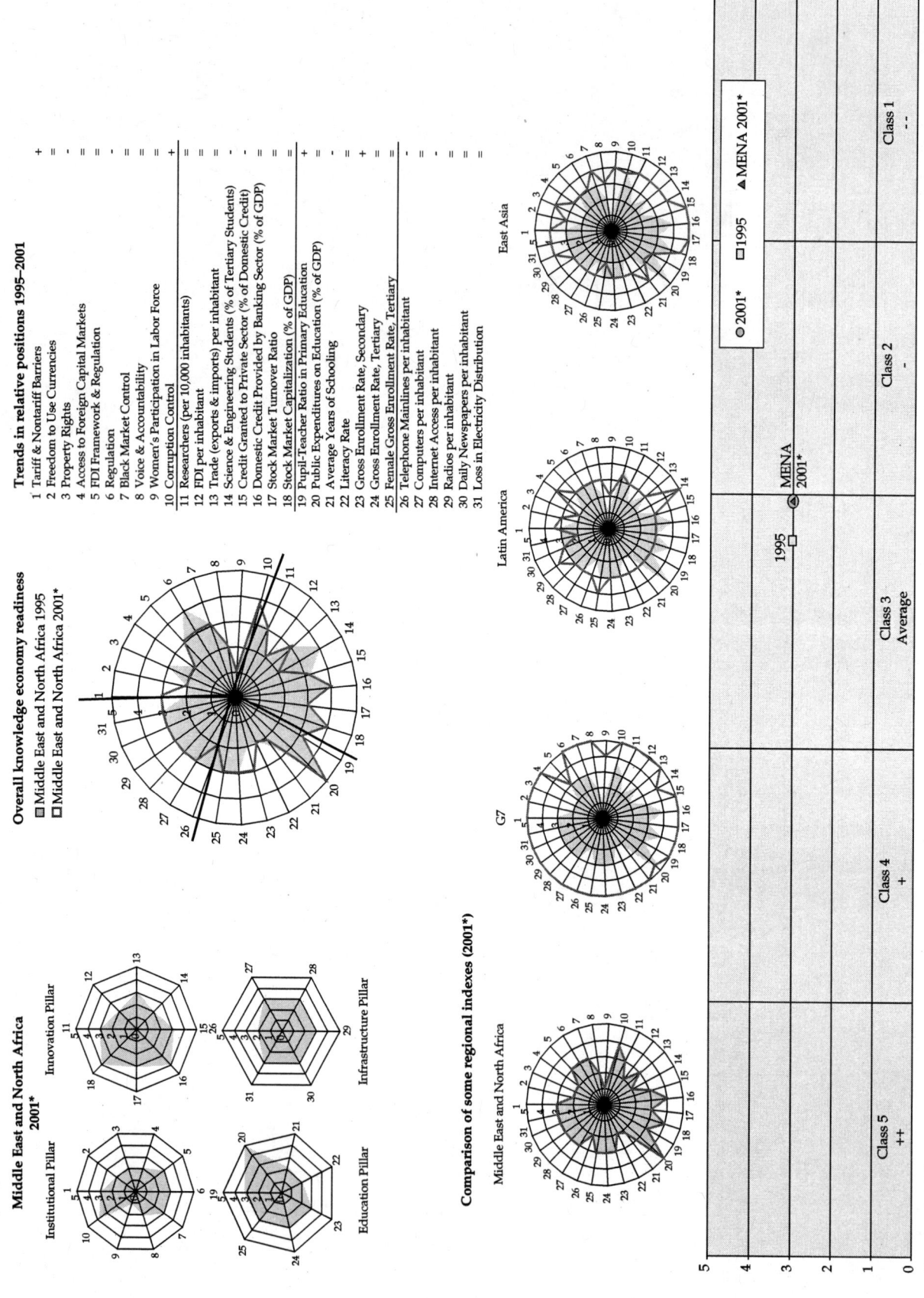

Overall knowledge economy readiness

▨ Middle East and North Africa 1995
▢ Middle East and North Africa 2001*

Trends in relative positions 1995–2001

1 Tariff & Nontariff Barriers — +
2 Freedom to Use Currencies — -
3 Property Rights — =
4 Access to Foreign Capital Markets — =
5 FDI Framework & Regulation — =
6 Regulation — ·
7 Black Market Control — =
8 Voice & Accountability — =
9 Women's Participation in Labor Force — =
10 Corruption Control — +
11 Researchers (per 10,000 inhabitants) — =
12 FDI per inhabitant — =
13 Trade (exports & imports) per inhabitant — =
14 Science & Engineering Students (% of Tertiary Students) — ·
15 Credit Granted to Private Sector (% of Domestic Credit) — =
16 Domestic Credit Provided by Banking Sector (% of GDP) — =
17 Stock Market Turnover Ratio — =
18 Stock Market Capitalization (% of GDP) — =
19 Pupil-Teacher Ratio in Primary Education — +
20 Public Expenditures on Education (% of GDP) — =
21 Average Years of Schooling — =
22 Literacy Rate — =
23 Gross Enrollment Rate, Secondary — +
24 Gross Enrollment Rate, Tertiary — =
25 Female Gross Enrollment Rate, Tertiary — =
26 Telephone Mainlines per inhabitant — ·
27 Computers per inhabitant — =
28 Internet Access per inhabitant — ·
29 Radios per inhabitant — ·
30 Daily Newspapers per inhabitant — =
31 Loss in Electricity Distribution — =

Middle East and North Africa 2001*

Institutional Pillar

Innovation Pillar

Education Pillar

Infrastructure Pillar

Comparison of some regional indexes (2001*)

Middle East and North Africa

G7

Latin America

East Asia

◎ 2001*   ▢ 1995   ▲ MENA 2001*

Class 1
Class 2
Class 3 Average
Class 4 +
Class 5 ++

*Or more recent data.

68

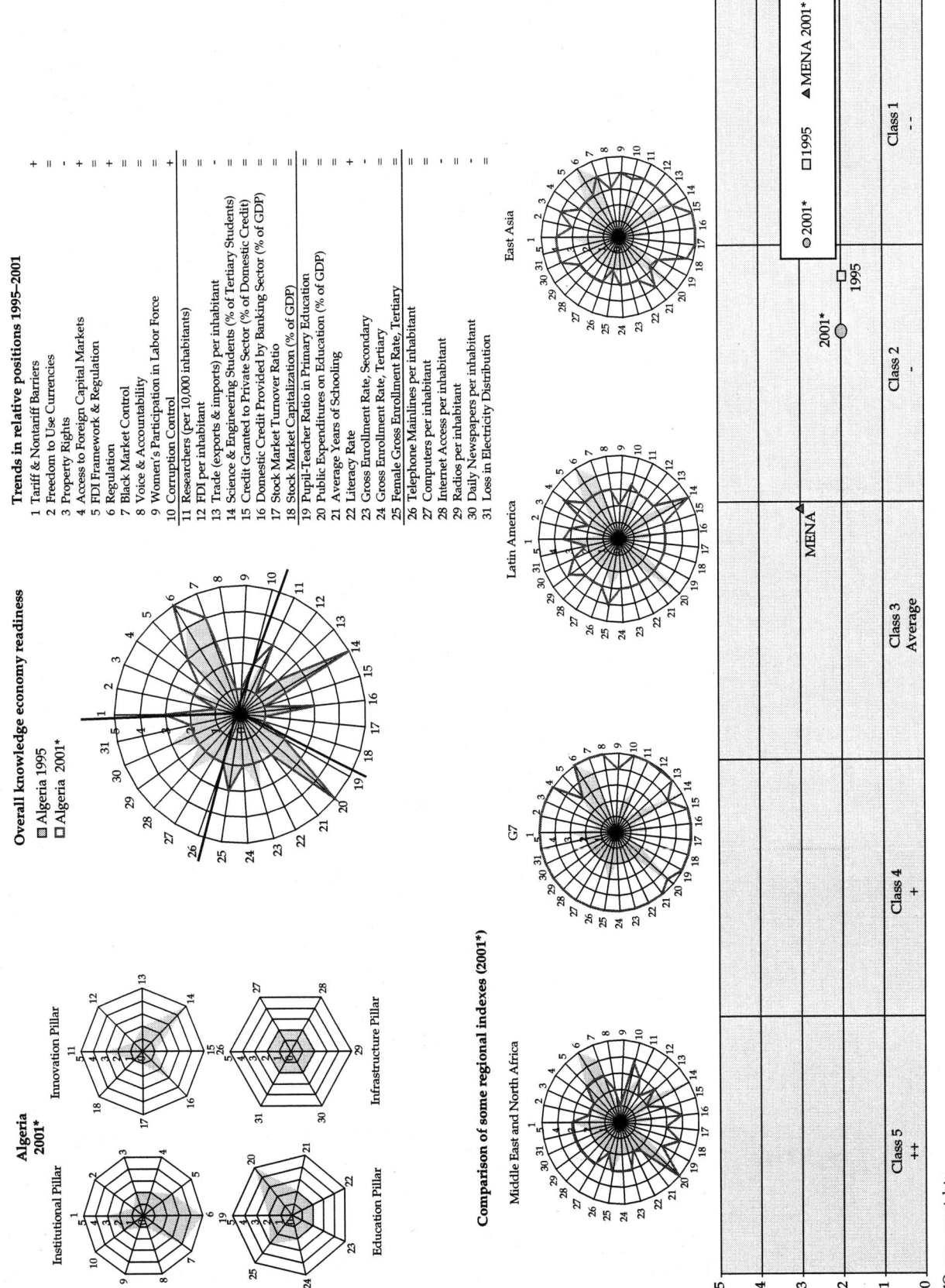

Trends in relative positions 1995–2001

1 Tariff & Nontariff Barriers — +
2 Freedom to Use Currencies — =
3 Property Rights — -
4 Access to Foreign Capital Markets — +
5 FDI Framework & Regulation — +
6 Regulation — +
7 Black Market Control — =
8 Voice & Accountability — =
9 Women's Participation in Labor Force — =
10 Corruption Control — +

11 Researchers (per 10,000 inhabitants) — =
12 FDI per inhabitant — -
13 Trade (exports & imports) per inhabitant — =
14 Science & Engineering Students (% of Tertiary Students) — =
15 Credit Granted to Private Sector (% of Domestic Credit) — =
16 Domestic Credit Provided by Banking Sector (% of GDP) — =
17 Stock Market Turnover Ratio — =
18 Stock Market Capitalization (% of GDP) — =
19 Pupil-Teacher Ratio in Primary Education — =
20 Public Expenditures on Education (% of GDP) — =
21 Average Years of Schooling — =
22 Literacy Rate — +
23 Gross Enrollment Rate, Secondary — -
24 Gross Enrollment Rate, Tertiary — =
25 Female Gross Enrollment Rate, Tertiary — =
26 Telephone Mainlines per inhabitant — =
27 Computers per inhabitant — =
28 Internet Access per inhabitant — -
29 Radios per inhabitant — -
30 Daily Newspapers per inhabitant — =
31 Loss in Electricity Distribution — =

Overall knowledge economy readiness

■ Algeria 1995
□ Algeria 2001*

Algeria 2001*

Institutional Pillar

Innovation Pillar

Education Pillar

Infrastructure Pillar

Comparison of some regional indexes (2001*)

Middle East and North Africa

G7

Latin America

East Asia

◎ 2001*   □ 1995   ▲ MENA 2001*

Class 5 ++   Class 4 +   Class 3 Average   Class 2 -   Class 1 --

*Or more recent data.

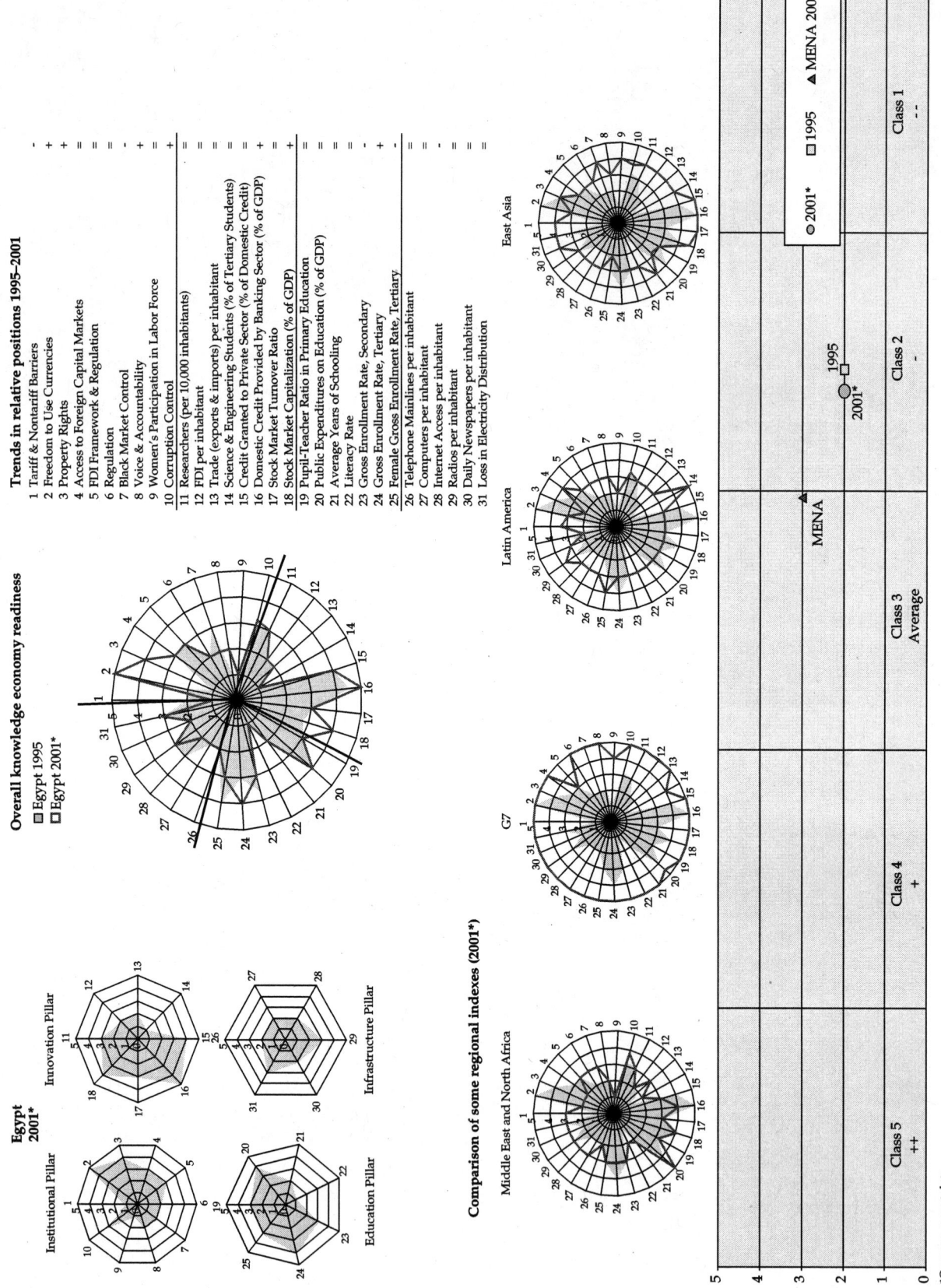

Trends in relative positions 1995–2001

1	Tariff & Nontariff Barriers	−
2	Freedom to Use Currencies	+
3	Property Rights	+
4	Access to Foreign Capital Markets	=
5	FDI Framework & Regulation	=
6	Regulation	−
7	Black Market Control	−
8	Voice & Accountability	+
9	Women's Participation in Labor Force	+
10	Corruption Control	+
11	Researchers (per 10,000 inhabitants)	=
12	FDI per inhabitant	=
13	Trade (exports & imports) per inhabitant	=
14	Science & Engineering Students (% of Tertiary Students)	=
15	Credit Granted to Private Sector (% of Domestic Credit)	=
16	Domestic Credit Provided by Banking Sector (% of GDP)	+
17	Stock Market Turnover Ratio	=
18	Stock Market Capitalization (% of GDP)	+
19	Pupil-Teacher Ratio in Primary Education	=
20	Public Expenditures on Education (% of GDP)	=
21	Average Years of Schooling	=
22	Literacy Rate	=
23	Gross Enrollment Rate, Secondary	−
24	Gross Enrollment Rate, Tertiary	=
25	Female Gross Enrollment Rate, Tertiary	+
26	Telephone Mainlines per inhabitant	=
27	Computers per inhabitant	=
28	Internet Access per inhabitant	−
29	Radios per inhabitant	=
30	Daily Newspapers per inhabitant	=
31	Loss in Electricity Distribution	=

Overall knowledge economy readiness

▨ Egypt 1995
▢ Egypt 2001*

Egypt 2001*

Innovation Pillar

Institutional Pillar

Infrastructure Pillar

Education Pillar

Comparison of some regional indexes (2001*)

Middle East and North Africa

G7

Latin America

East Asia

◉ 2001*   ▢ 1995   ▲ MENA 2001*

Class 5 ++   Class 4 +   Class 3 Average   Class 2 −   Class 1 − −

*Or more recent data.

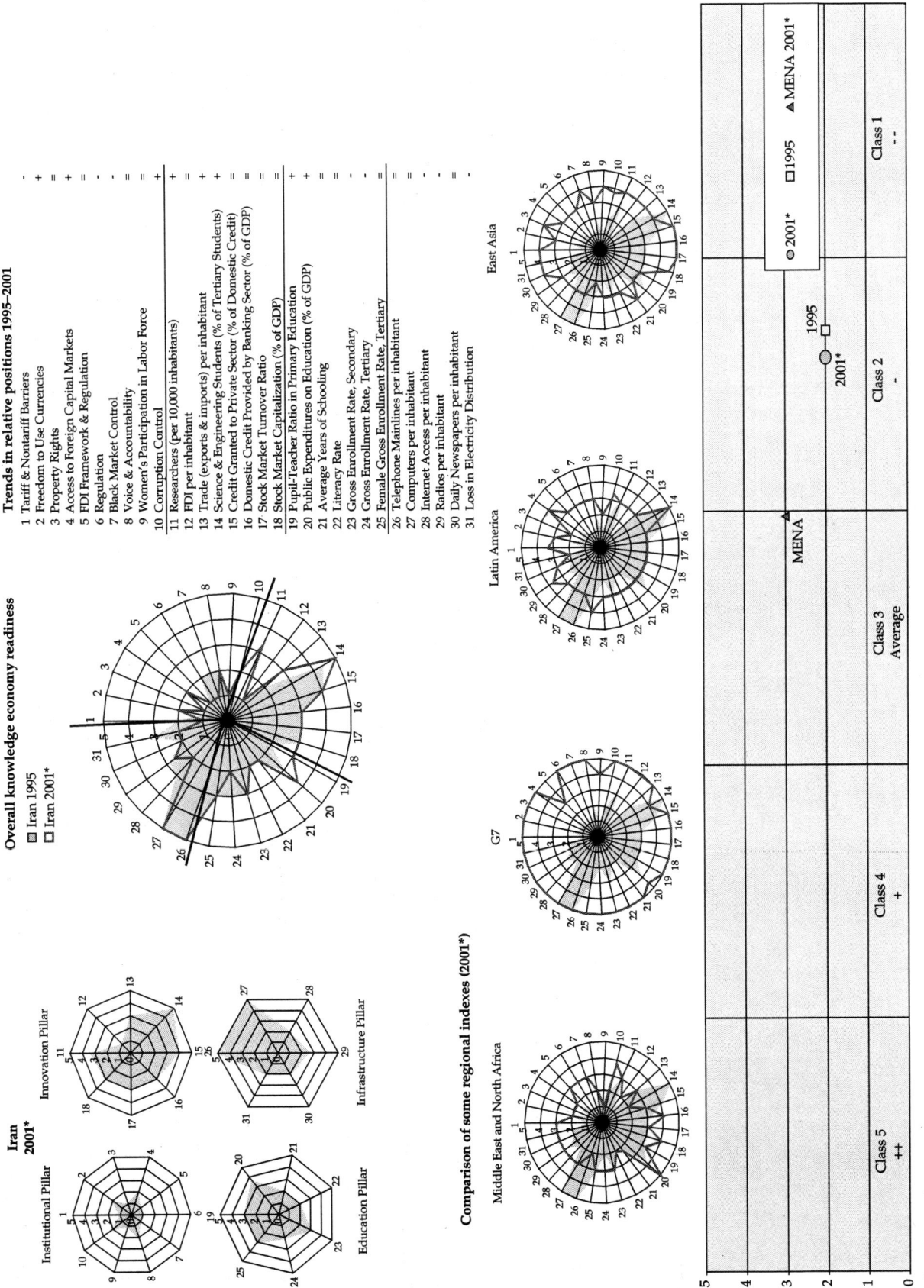

Trends in relative positions 1995–2001

1	Tariff & Nontariff Barriers	-
2	Freedom to Use Currencies	+
3	Property Rights	=
4	Access to Foreign Capital Markets	+
5	FDI Framework & Regulation	+
6	Regulation	-
7	Black Market Control	=
8	Voice & Accountability	=
9	Women's Participation in Labor Force	=
10	Corruption Control	+
11	Researchers (per 10,000 inhabitants)	=
12	FDI per inhabitant	=
13	Trade (exports & imports) per inhabitant	+
14	Science & Engineering Students (% of Tertiary Students)	+
15	Credit Granted to Private Sector (% of Domestic Credit)	=
16	Domestic Credit Provided by Banking Sector (% of GDP)	=
17	Stock Market Turnover Ratio	=
18	Stock Market Capitalization (% of GDP)	=
19	Pupil-Teacher Ratio in Primary Education	+
20	Public Expenditures on Education (% of GDP)	+
21	Average Years of Schooling	=
22	Literacy Rate	-
23	Gross Enrollment Rate, Secondary	=
24	Gross Enrollment Rate, Tertiary	=
25	Female Gross Enrollment Rate, Tertiary	=
26	Telephone Mainlines per inhabitant	=
27	Computers per inhabitant	-
28	Internet Access per inhabitant	-
29	Radios per inhabitant	-
30	Daily Newspapers per inhabitant	=
31	Loss in Electricity Distribution	-

Overall knowledge economy readiness

☑ Iran 1995
☐ Iran 2001*

Iran 2001*

Institutional Pillar

Innovation Pillar

Infrastructure Pillar

Education Pillar

Comparison of some regional indexes (2001*)

Middle East and North Africa

G7

Latin America

East Asia

◎ 2001*   ☐ 1995   ▲ MENA 2001*

Class 5	Class 4	Class 3	Class 2	Class 1
++	+	Average	-	--

*Or more recent data.

71

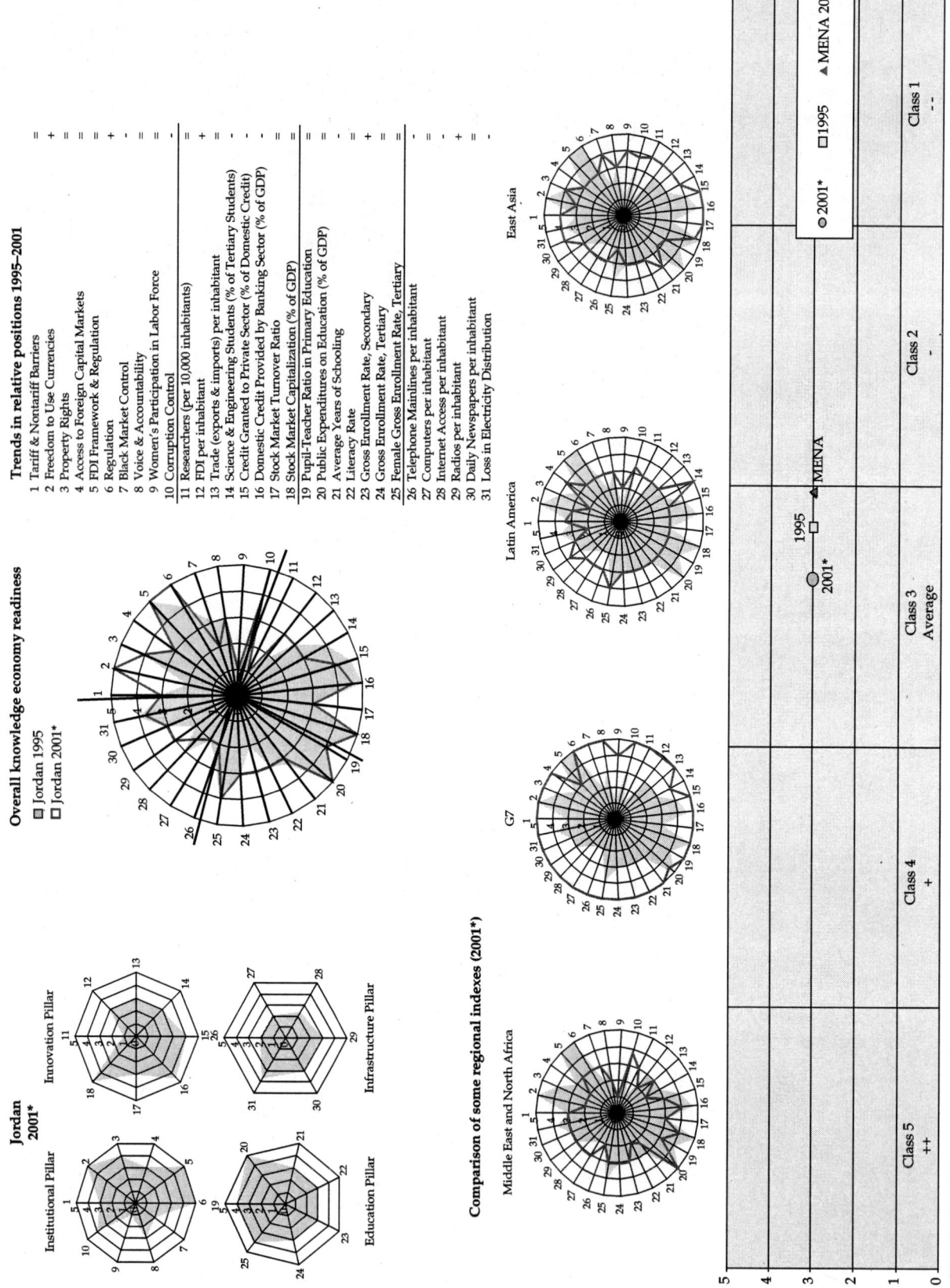

Trends in relative positions 1995–2001

1 Tariff & Nontariff Barriers — =
2 Freedom to Use Currencies — +
3 Property Rights — =
4 Access to Foreign Capital Markets — =
5 FDI Framework & Regulation — =
6 Regulation — +
7 Black Market Control — ·
8 Voice & Accountability — =
9 Women's Participation in Labor Force — ·
10 Corruption Control — ·
11 Researchers (per 10,000 inhabitants) — =
12 FDI per inhabitant — +
13 Trade (exports & imports) per inhabitant — ·
14 Science & Engineering Students (% of Tertiary Students) — ·
15 Credit Granted to Private Sector (% of Domestic Credit) — ·
16 Domestic Credit Provided by Banking Sector (% of GDP) — ·
17 Stock Market Turnover Ratio — ·
18 Stock Market Capitalization (% of GDP) — =
19 Pupil-Teacher Ratio in Primary Education — =
20 Public Expenditures on Education (% of GDP) — =
21 Average Years of Schooling — ·
22 Literacy Rate — +
23 Gross Enrollment Rate, Secondary — =
24 Gross Enrollment Rate, Tertiary — =
25 Female Gross Enrollment Rate, Tertiary — =
26 Telephone Mainlines per inhabitant — =
27 Computers per inhabitant — =
28 Internet Access per inhabitant — =
29 Radios per inhabitant — +
30 Daily Newspapers per inhabitant — =
31 Loss in Electricity Distribution — ·

Overall knowledge economy readiness

☐ Jordan 1995
☐ Jordan 2001*

Jordan 2001*

Institutional Pillar

Innovation Pillar

Education Pillar

Infrastructure Pillar

Comparison of some regional indexes (2001*)

Middle East and North Africa

G7

Latin America

East Asia

⊚ 2001*   ☐ 1995   ▲ MENA 2001*

Class 5 ++   Class 4 +   Class 3 Average   Class 2 –   Class 1 – –

*Or more recent data.

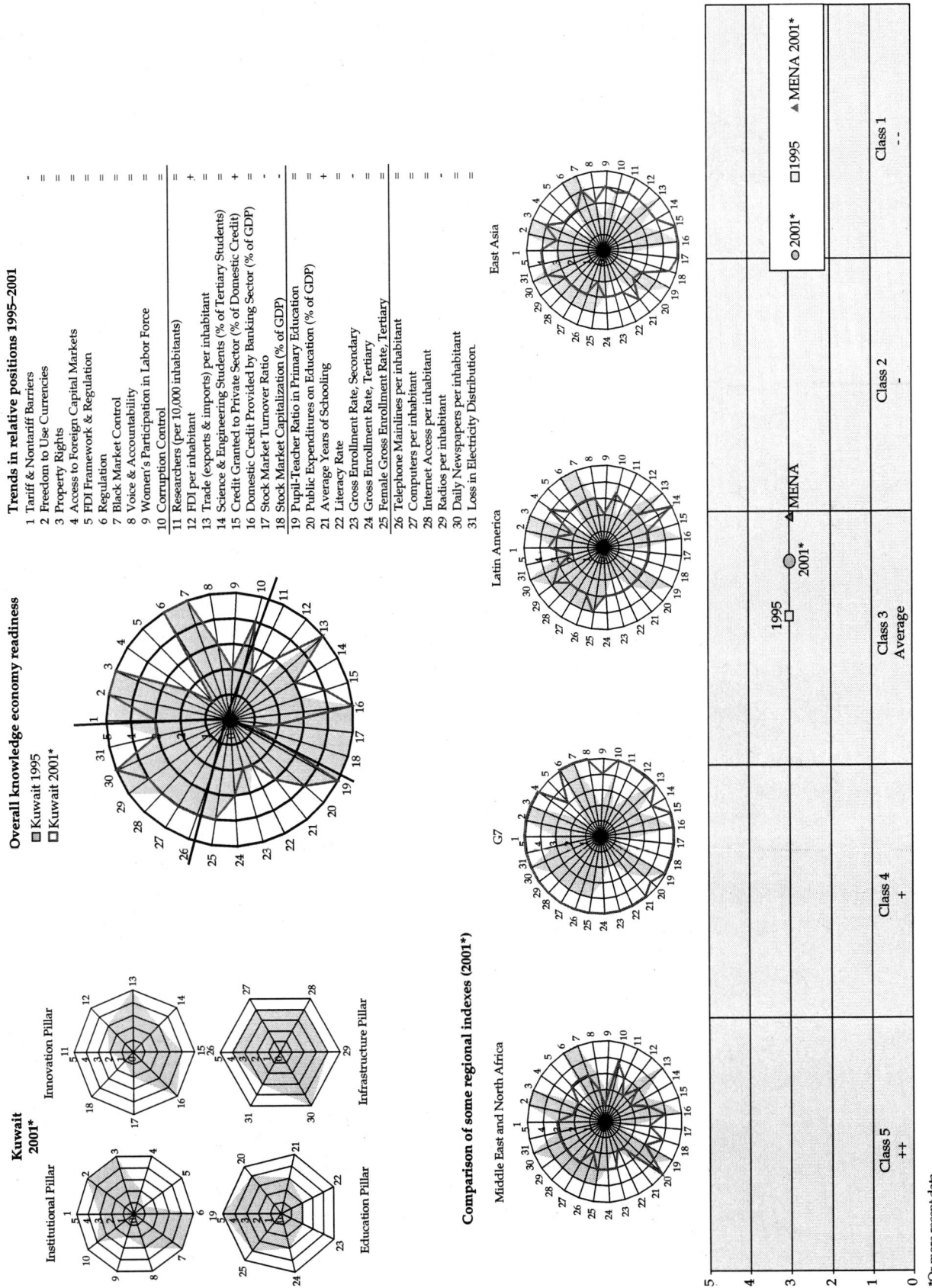

**Kuwait
2001***

Institutional Pillar

Innovation Pillar

Education Pillar

Infrastructure Pillar

Overall knowledge economy readiness

◼ Kuwait 1995
☐ Kuwait 2001*

Trends in relative positions 1995–2001

1 Tariff & Nontariff Barriers	-
2 Freedom to Use Currencies	=
3 Property Rights	=
4 Access to Foreign Capital Markets	=
5 FDI Framework & Regulation	=
6 Regulation	=
7 Black Market Control	=
8 Voice & Accountability	=
9 Women's Participation in Labor Force	=
10 Corruption Control	=
11 Researchers (per 10,000 inhabitants)	-
12 FDI per inhabitant	+
13 Trade (exports & imports) per inhabitant	=
14 Science & Engineering Students (% of Tertiary Students)	=
15 Credit Granted to Private Sector (% of Domestic Credit)	+
16 Domestic Credit Provided by Banking Sector (% of GDP)	=
17 Stock Market Turnover Ratio	-
18 Stock Market Capitalization (% of GDP)	-
19 Pupil-Teacher Ratio in Primary Education	=
20 Public Expenditures on Education (% of GDP)	+
21 Average Years of Schooling	-
22 Literacy Rate	=
23 Gross Enrollment Rate, Secondary	-
24 Gross Enrollment Rate, Tertiary	=
25 Female Gross Enrollment Rate, Tertiary	=
26 Telephone Mainlines per inhabitant	=
27 Computers per inhabitant	=
28 Internet Access per inhabitant	-
29 Radios per inhabitant	=
30 Daily Newspapers per inhabitant	=
31 Loss in Electricity Distribution.	=

Comparison of some regional indexes (2001*)

Middle East and North Africa

G7

Latin America

East Asia

◎ 2001*   ☐ 1995   ▲ MENA 2001*

Class 1
- -

Class 2
-

Class 3
Average

Class 4
+

Class 5
++

1995 ☐   2001* ◎   ▲ MENA

*Or more recent data.

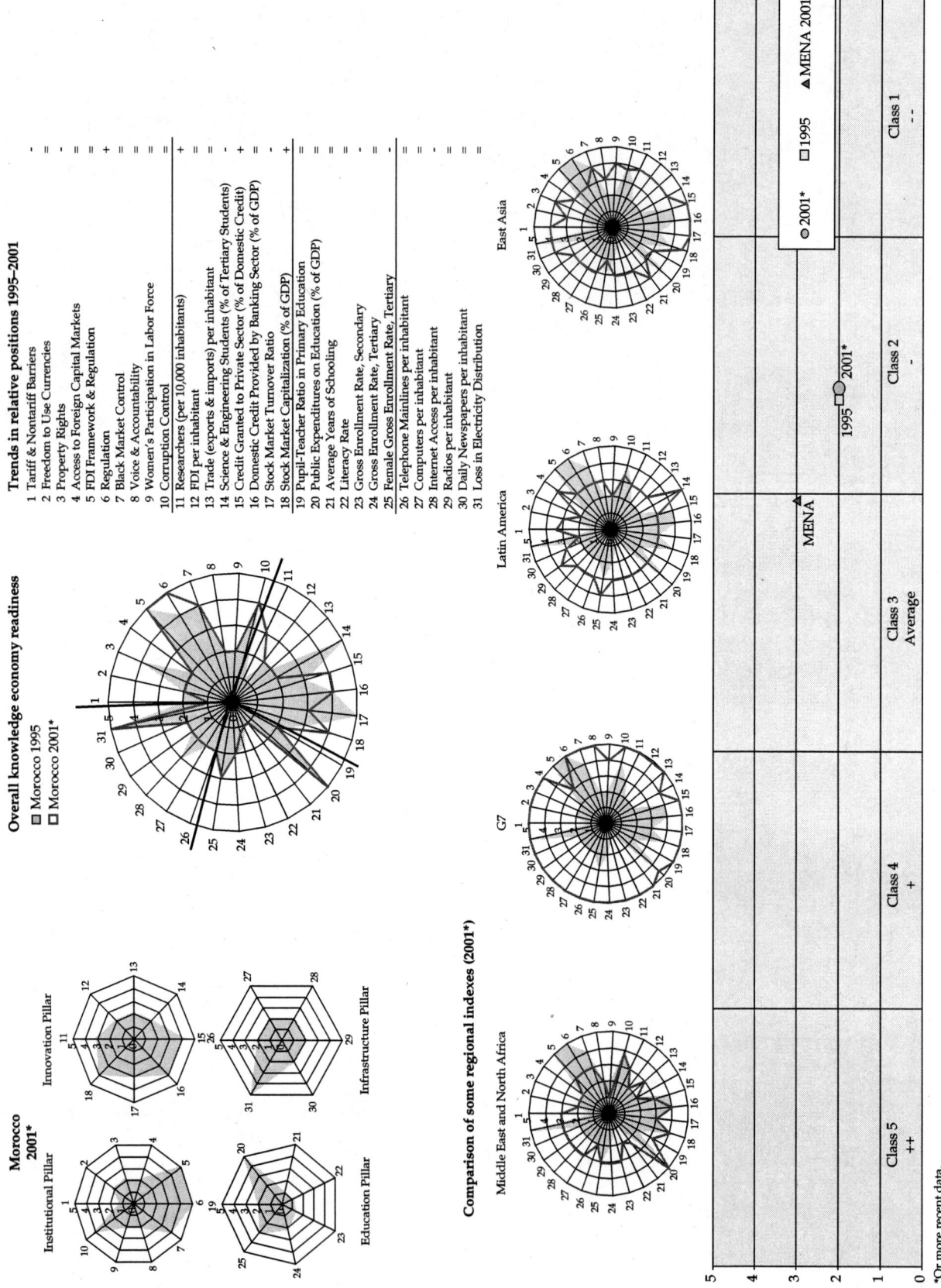

Overall knowledge economy readiness

■ Morocco 1995
□ Morocco 2001*

Morocco 2001*

Institutional Pillar

Innovation Pillar

Education Pillar

Infrastructure Pillar

Trends in relative positions 1995–2001

1 Tariff & Nontariff Barriers — =
2 Freedom to Use Currencies — =
3 Property Rights =
4 Access to Foreign Capital Markets =
5 FDI Framework & Regulation +
6 Regulation +
7 Black Market Control =
8 Voice & Accountability =
9 Women's Participation in Labor Force =
10 Corruption Control =

11 Researchers (per 10,000 inhabitants) +
12 FDI per inhabitant +
13 Trade (exports & imports) per inhabitant =
14 Science & Engineering Students (% of Tertiary Students) +
15 Credit Granted to Private Sector (% of Domestic Credit) –
16 Domestic Credit Provided by Banking Sector (% of GDP) +
17 Stock Market Turnover Ratio =
18 Stock Market Capitalization (% of GDP) +

19 Pupil-Teacher Ratio in Primary Education =
20 Public Expenditures on Education (% of GDP) =
21 Average Years of Schooling =
22 Literacy Rate –
23 Gross Enrollment Rate, Secondary –
24 Gross Enrollment Rate, Tertiary –
25 Female Gross Enrollment Rate, Tertiary –

26 Telephone Mainlines per inhabitant =
27 Computers per inhabitant +
28 Internet Access per inhabitant –
29 Radios per inhabitant –
30 Daily Newspapers per inhabitant =
31 Loss in Electricity Distribution =

Comparison of some regional indexes (2001*)

Middle East and North Africa

G7

Latin America

East Asia

○ 2001*   □ 1995   ▲ MENA 2001*

Class 5	Class 4	Class 3	Class 2	Class 1
++	+	Average	–	– –

MENA   1995 □ ○ 2001*

*Or more recent data.

74

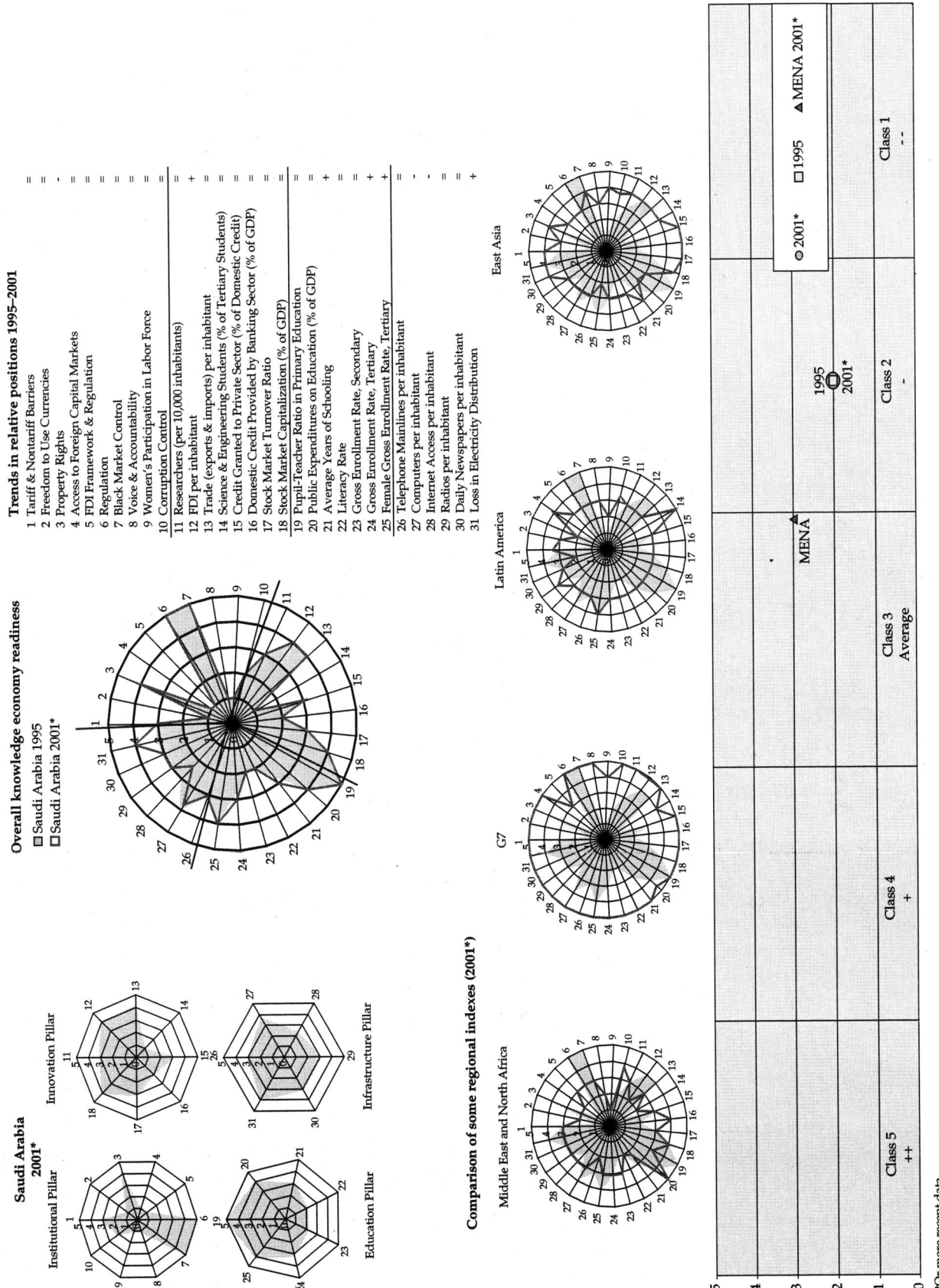

Trends in relative positions 1995–2001

1 Tariff & Nontariff Barriers
2 Freedom to Use Currencies
3 Property Rights
4 Access to Foreign Capital Markets
5 FDI Framework & Regulation
6 Regulation
7 Black Market Control
8 Voice & Accountability
9 Women's Participation in Labor Force
10 Corruption Control
11 Researchers (per 10,000 inhabitants)
12 FDI per inhabitant
13 Trade (exports & imports) per inhabitant
14 Science & Engineering Students (% of Tertiary Students)
15 Credit Granted to Private Sector (% of Domestic Credit)
16 Domestic Credit Provided by Banking Sector (% of GDP)
17 Stock Market Turnover Ratio
18 Stock Market Capitalization (% of GDP)
19 Pupil-Teacher Ratio in Primary Education
20 Public Expenditures on Education (% of GDP)
21 Average Years of Schooling
22 Literacy Rate
23 Gross Enrollment Rate, Secondary
24 Gross Enrollment Rate, Tertiary
25 Female Gross Enrollment Rate, Tertiary
26 Telephone Mainlines per inhabitant
27 Computers per inhabitant
28 Internet Access per inhabitant
29 Radios per inhabitant
30 Daily Newspapers per inhabitant
31 Loss in Electricity Distribution

Overall knowledge economy readiness
Saudi Arabia 1995
Saudi Arabia 2001*

Saudi Arabia 2001*

Institutional Pillar
Innovation Pillar
Education Pillar
Infrastructure Pillar

Comparison of some regional indexes (2001*)

Middle East and North Africa
G7
Latin America
East Asia

2001*   1995   MENA 2001*

Class 1 --
Class 2 -
Class 3 Average
Class 4 +
Class 5 ++

MENA

*Or more recent data.

75

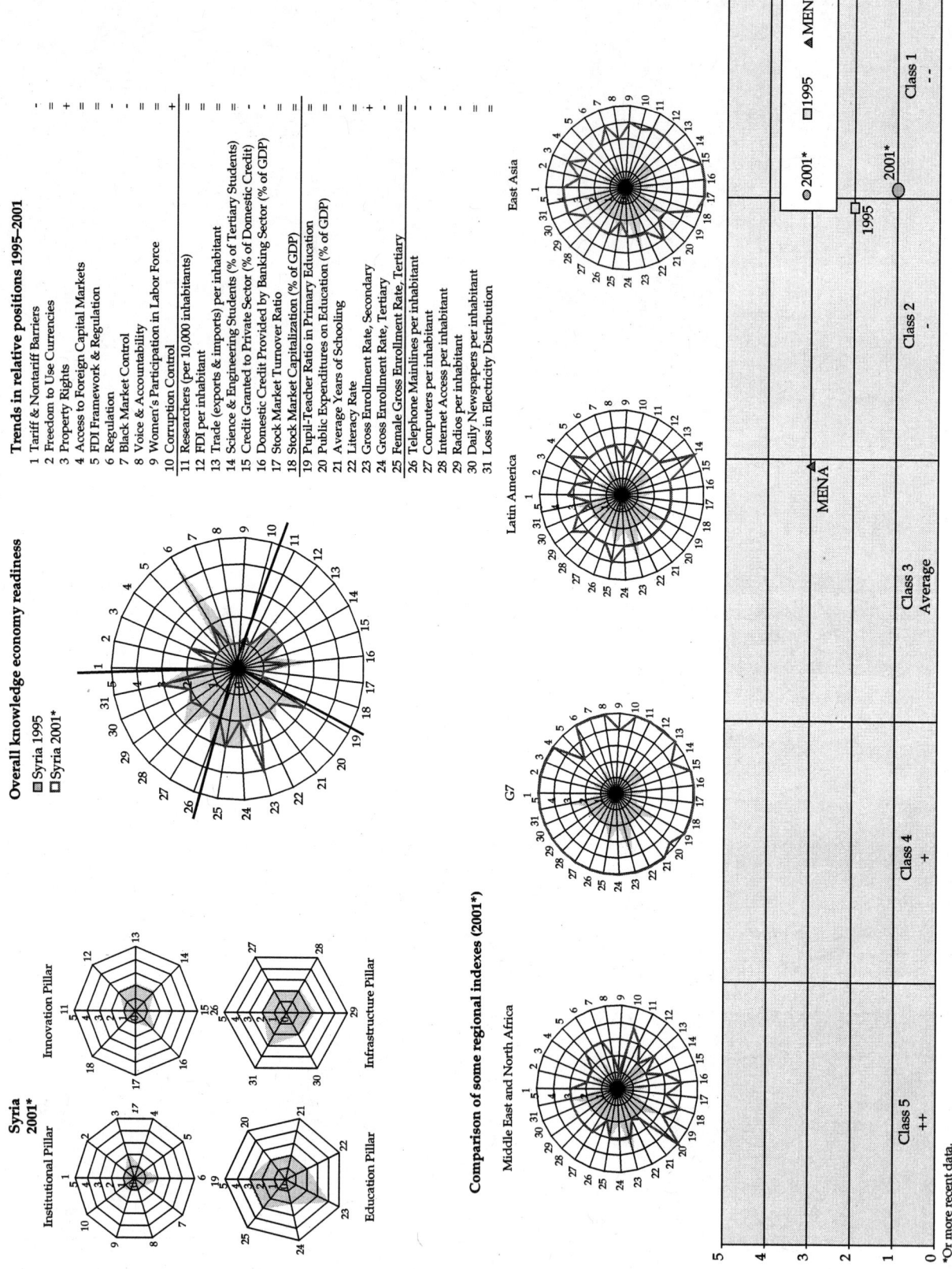

Overall knowledge economy readiness

- Syria 1995
- Syria 2001*

Syria 2001*

Institutional Pillar

Innovation Pillar

Education Pillar

Infrastructure Pillar

Trends in relative positions 1995–2001

1 Tariff & Nontariff Barriers —
2 Freedom to Use Currencies =
3 Property Rights +
4 Access to Foreign Capital Markets =
5 FDI Framework & Regulation =
6 Regulation ·
7 Black Market Control ·
8 Voice & Accountability =
9 Women's Participation in Labor Force =
10 Corruption Control +
11 Researchers (per 10,000 inhabitants) =
12 FDI per inhabitant =
13 Trade (exports & imports) per inhabitant =
14 Science & Engineering Students (% of Tertiary Students) =
15 Credit Granted to Private Sector (% of Domestic Credit) ·
16 Domestic Credit Provided by Banking Sector (% of GDP) ·
17 Stock Market Turnover Ratio =
18 Stock Market Capitalization (% of GDP) =
19 Pupil-Teacher Ratio in Primary Education =
20 Public Expenditures on Education (% of GDP) =
21 Average Years of Schooling =
22 Literacy Rate +
23 Gross Enrollment Rate, Secondary =
24 Gross Enrollment Rate, Tertiary ·
25 Female Gross Enrollment Rate, Tertiary ·
26 Telephone Mainlines per inhabitant ·
27 Computers per inhabitant ·
28 Internet Access per inhabitant ·
29 Radios per inhabitant =
30 Daily Newspapers per inhabitant ·
31 Loss in Electricity Distribution =

Comparison of some regional indexes (2001*)

Middle East and North Africa

G7

Latin America

East Asia

- 2001*
- 1995
- MENA 2001*

Class 5 ++   Class 4 +   Class 3 Average   Class 2 -   Class 1 - -

*Or more recent data.

76

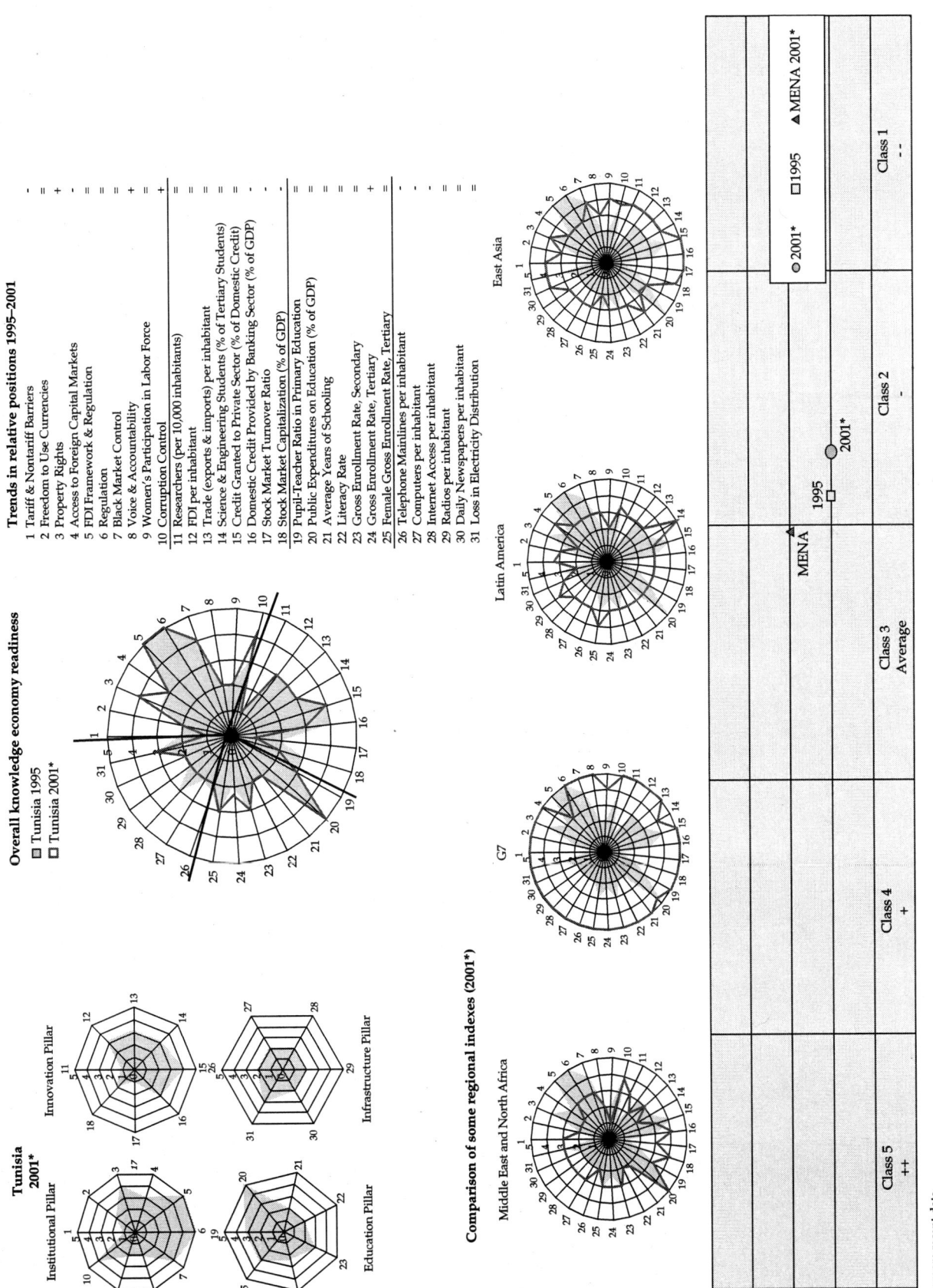

Overall knowledge economy readiness

■ Tunisia 1995
□ Tunisia 2001*

Tunisia 2001*

Institutional Pillar

Innovation Pillar

Education Pillar

Infrastructure Pillar

Comparison of some regional indexes (2001*)

Middle East and North Africa

G7

Latin America

East Asia

Trends in relative positions 1995–2001

1 Tariff & Nontariff Barriers — -
2 Freedom to Use Currencies — =
3 Property Rights — +
4 Access to Foreign Capital Markets — -
5 FDI Framework & Regulation — =
6 Regulation — =
7 Black Market Control — =
8 Voice & Accountability — +
9 Women's Participation in Labor Force — =
10 Corruption Control — +

11 Researchers (per 10,000 inhabitants) — =
12 FDI per inhabitant — =
13 Trade (exports & imports) per inhabitant — =
14 Science & Engineering Students (% of Tertiary Students) — =
15 Credit Granted to Private Sector (% of Domestic Credit) — =
16 Domestic Credit Provided by Banking Sector (% of GDP) — -
17 Stock Market Turnover Ratio — -
18 Stock Market Capitalization (% of GDP) — =

19 Pupil-Teacher Ratio in Primary Education — =
20 Public Expenditures on Education (% of GDP) — =
21 Average Years of Schooling — =
22 Literacy Rate — =
23 Gross Enrollment Rate, Secondary — =
24 Gross Enrollment Rate, Tertiary — +
25 Female Gross Enrollment Rate, Tertiary — =

26 Telephone Mainlines per inhabitant — -
27 Computers per inhabitant — -
28 Internet Access per inhabitant — -
29 Radios per inhabitant — =
30 Daily Newspapers per inhabitant — =
31 Loss in Electricity Distribution — =

◎ 2001*   □ 1995   ▲ MENA 2001*

Class 1
– –

Class 2
–

Class 3
Average

Class 4
+

Class 5
++

MENA

1995

2001*

*Or more recent data.

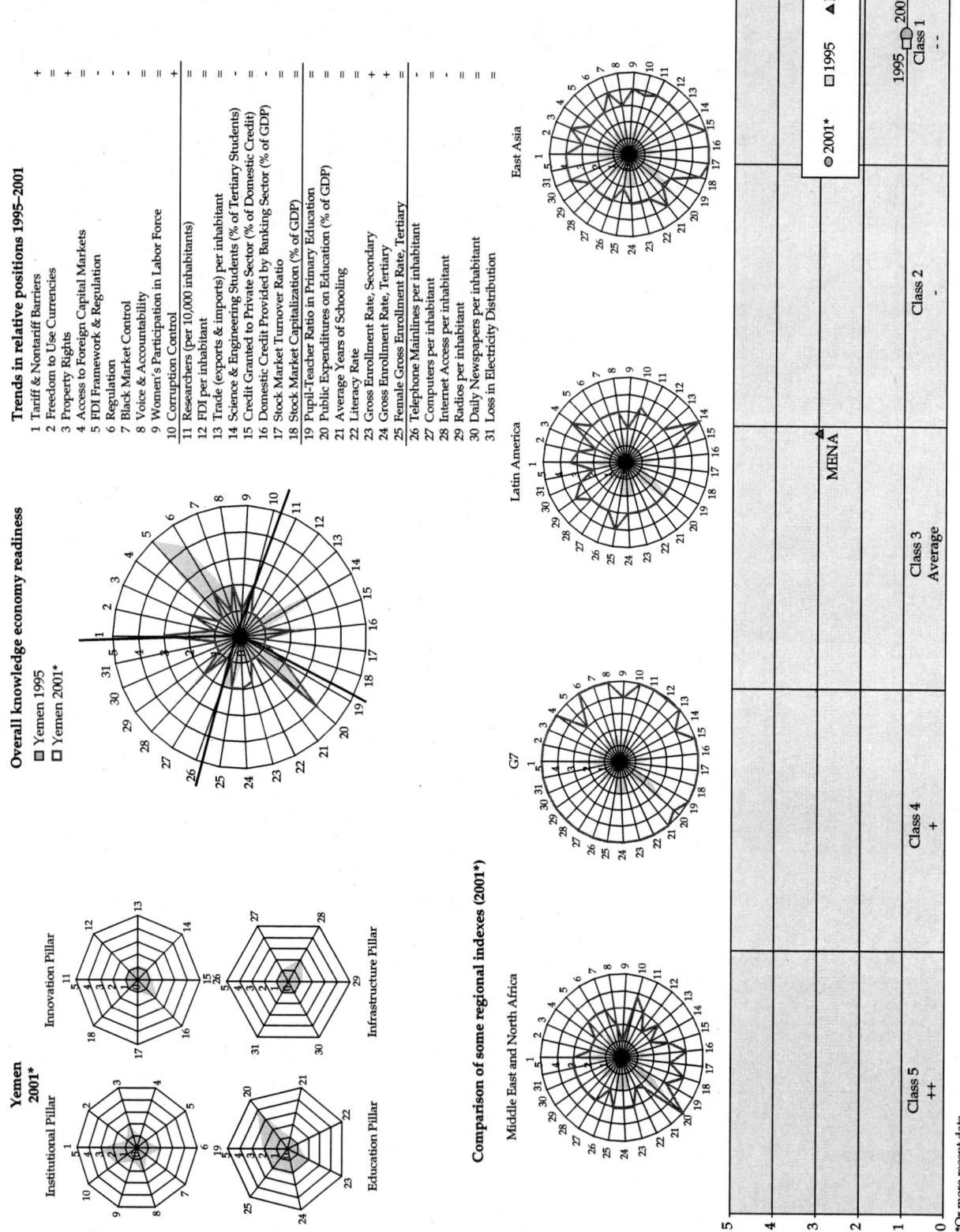

Overall knowledge economy readiness

- ☐ Yemen 1995
- ☐ Yemen 2001*

Yemen 2001*

Institutional Pillar

Innovation Pillar

Education Pillar

Infrastructure Pillar

Trends in relative positions 1995–2001

1	Tariff & Nontariff Barriers	+
2	Freedom to Use Currencies	=
3	Property Rights	+
4	Access to Foreign Capital Markets	=
5	FDI Framework & Regulation	-
6	Regulation	-
7	Black Market Control	-
8	Voice & Accountability	=
9	Women's Participation in Labor Force	=
10	Corruption Control	+
11	Researchers (per 10,000 inhabitants)	=
12	FDI per inhabitant	=
13	Trade (exports & imports) per inhabitant	=
14	Science & Engineering Students (% of Tertiary Students)	-
15	Credit Granted to Private Sector (% of Domestic Credit)	=
16	Domestic Credit Provided by Banking Sector (% of GDP)	=
17	Stock Market Turnover Ratio	=
18	Stock Market Capitalization (% of GDP)	=
19	Pupil-Teacher Ratio in Primary Education	=
20	Public Expenditures on Education (% of GDP)	=
21	Average Years of Schooling	=
22	Literacy Rate	=
23	Gross Enrollment Rate, Secondary	+
24	Gross Enrollment Rate, Tertiary	+
25	Female Gross Enrollment Rate, Tertiary	=
26	Telephone Mainlines per inhabitant	-
27	Computers per inhabitant	-
28	Internet Access per inhabitant	-
29	Radios per inhabitant	=
30	Daily Newspapers per inhabitant	=
31	Loss in Electricity Distribution	=

Comparison of some regional indexes (2001*)

Middle East and North Africa

G7

Latin America

East Asia

- ◎ 2001*
- ☐ 1995
- ▲ MENA 2001*

1995 ☐☐ 2001*

| Class 5 | Class 4 | Class 3 | Class 2 | Class 1 |
| ++ | + | Average | | |

MENA

*Or more recent data.

78

References

The word *processed* describes informally reproduced works that may not be commonly available through libraries.

Benmokthar, Rachid. 2002. "Knowledge for Development: The Case of Morocco." Presentation at the Knowledge for Development Forum, Marseilles. Available at: www.worldbank.org/k4dmarseille.

Bensaid, Moncef. 2002. "Le savoir et le savoir-faire au service du développement. L'exemple de la réforme tunisienne de la formation professionnelle." Presentation at the Knowledge for Development Forum, Marseilles. Available at: www.worldbank.org/k4dmarseille.

Blanchard, Olivier. 1997. *Macroeconomics.* New York: Prentice Hall.

Boutrolle, Clotilde. 2002. "Economie de la connaissance: une étude des pays MENA." Consultant report. World Bank, MENA Region. Processed.

Dahlman, Carl. 2002. "Challenges for MENA Countries from the Knowledge Revolution." Washington, D.C.: World Bank. Available at: www.worldbank.org/k4dmarseille under "Benchmarking MENA Countries Readiness for KE."

Djeflat, Abdelkader. 2002. "National Systems of Innovation in the MENA Region." Washington, D.C.: World Bank.

Easterly, William, and Ross Levine. 2001. "What Have We Learned from a Decade of Empirical Research on Growth? It's Not Factor Accumulation: Stylized Facts and Growth Models." *World Bank Economic Review* 15: 177.

Eliasson, Gunnar. 1996. *Firms' Objectives, Control and Organisation: The Use of Information and the Transfer of Knowledge within a Firm.* Dordrecht, Boston, and London: Kluwer Academic Publishers.

Gago, Jose M. 2002. "Portugal as a Test-Bed for Democracy and Economic Development." Presentation at the Knowledge for Development Forum, Marseilles. Available at: www.worldbank.org/k4dmarseille.

Gertler, P., and G. Blalock. 2001. "Technology Acquisition in Indonesian Manufacturing: The Effects of Foreign Direct Investment and Exports." Paper presented at the conference on East Asia's Future Economy, October, World Bank and the Kennedy School of Government, Harvard University

Hoekman, Bernard, and Patrick Messerlin. 2002. "Harnessing Trade for Development and Growth in the Middle East." Washington, D.C.: Council on Foreign Relations.

John, K. J. 2002. "Malaysia's GEM: Leapfrogging into the K-E Economy via ICT Enablement." Presentation at the Knowledge for Development Forum, Marseilles. Available at: www.worldbank.org/k4dmarseille.

Kang, Young-Chul. 2002. "How MBN Has Promoted Knowledge Revolution in Korea." Presentation at the Knowledge for Development Forum, Marseilles. Available at: www.worldbank.org/k4dmarseille.

Keller, Jennifer. and Mustapha Nabli. 2002. "The Macroeconomics of Labor Market Outcomes in MENA over the 1990s: How Growth Has Failed to Keep Pace with a Burgeoning Labor Market." Working Paper no. 71. World Bank, Washington, D.C.

Lundvall, Bengt Ake. 1998. "The Learning Economy: Challenger to Economic Theory and Policy." In K. Nielsen and B. Johnson, eds., *Institutions and Economic Change.* Cheltenham, U.K.: Edward Elgar Publishers.

Maqusi, Mohammad. 2002. "Establishing an Open University with a Difference: The Case of the Arab Open University." Presentation at the Knowledge for Development Forum, Marseilles. Available at: www.worldbank.org/k4dmarseille.

Maskus, K. E., and M. Penubarti. 1995. "How Trade-Related Are Intellectual Property Rights?" *Journal of International Economics* 39: 227–48.

Mezouar, Abdelkebir, ed. 2002. *L'entreprise marocaine et la modernité.* Casablanca: Centre de Recherches des Dirigeants.

Organisation for Economic Co-operation and Development (OECD). 2000. *Managing Innovation Systems.* Paris.

Page, John. 2002. "Structural Reforms in the Middle East and North Africa." In Peter Cornelius and Klaus Schwab, eds., *Arab World Competitiveness Report 2002–2003.* Oxford, U.K.: Oxford University Press.

Rischard, Jean-François. 2002. "The Knowledge Economy—World-Wide Trends." Washington, D.C.: World Bank. Presentation at the Knowledge for Development Forum, Marseilles. Available at: www.worldbank.org/k4dmarseille.

Ruthven, Malise. 1997. *Islam: A Very Short Introduction.* Oxford, U.K.: Oxford University Press.

Saidi, Nasser. 2002. "Lebanon and Syria: Building a Strategic Partnership." Document communicated by the author. August 4.

Saidi, Nasser, and Hala Yared. 2002. "E-Government: Technology for Good Governance, Development and Democracy in the MENA Countries." Communication at Mediterranean Development Forum 4. Amman.

Sarbib, Jean-Louis. 2002. "Concluding Remarks." Knowledge for Development Forum, Marseilles. Available at : www.worldbank.org/k4dmarseille.

Serres, Michel. 2002. "L'humanisme universel qui vient." *Le Monde,* July 5.

United Nations Development Programme (UNDP). 2002. *Arab Human Development Report 2002.* New York and Oxford, U.K.: Oxford University Press.

Wilson, Rodney. 2002. "Arab Banking and Capital Market Developments." In World Economic Forum, *Arab World Competitiveness Report 2002–2003.* Peter Cornelius and Klaus Schwab, eds. Oxford: Oxford University Press.

World Bank. 1998. *World Development Report 1998–1999: Knowledge for Development.* Washington, D.C.

———. 2001. *Global Economic Prospects.* Washington, D.C.

———. 2002. "Closing the Skills and Technology Gap." Latin American and Caribbean Region, Washington, D.C.

World Economic Forum. 2002. In Peter Cornelius and Klaus Schwab, eds., *The Global Competitiveness Report 2002–2003.* Oxford, U.K.: Oxford University Press.

———. 2003. In Peter Cornelius and Klaus Schwab, eds., *Arab World Competitiveness Report 2002–2003.* Oxford, U.K.: Oxford University Press.

Yacoubian. M. 2002. "Dialogue Across Cultures." Summary of proceedings, May 2–3, World Bank, Washington, D.C.

Documents Presented at the World Bank Forum on Knowledge for Development in the Middle East and North Africa

September 9–12, 2002, Marseilles, France

These documents are available online at http://www.worldbank.org/k4dmarseille

1. Conference Program and Objectives
   - Objectives and Topics (also available in French)
   - Agenda
   - List of Participants
2. Knowledge Economy for MENA Countries: Why? How?
   - Introductory Note to the Conference (also available in French)
   - Arab Development Issues: UNDP Presentation (M. Doraid) and UNDP Arab Development Report
   - Knowledge Economy: World Wide Trends (J.-F. Rischard)
   - Benchmarking MENA Countries' Readiness for the Knowledge Economy (C. Dahlman)
3. Conference Background Reports
   - General Background Report (J.-L. Reiffers and J.-E. Aubert (also available in French)
   - Report on Innovation (A. Djeflat)
   - Report on Education (A. Kirchberger)
   - Report on Telecom Infrastructure Development for Corporate Data (Analysys Consulting)
4. Selected Country Assessments and Strategies
   - Dubai (O. Bin Sulaiman)
   - Jordan (B. Zu'bi and N. Fayoumi)
   - Korea (Y.C. Kang)
   - Malaysia (K.J. John)
   - Morocco (R. Benmokthar, R. Slimi, A. Mezouar, and A. Driouchi)
   - Portugal (M. Gago)
5. ICT Studies and Benchmarking
   - Benchmarking Arab Regulators (M.A. Mustafa)
   - Telecom Infrastructure Telecom Reform: The Moroccan Case (M. Terrab)
   - ICT at a Glance: Country Tables
   - ICT for Development in the Arab States (G. Accascina)
   - Women and ICT at: www.cddc.vt.edu/knownet/articles/womenandICT.htm
6. Virtual Universities and Regional Initiatives
   - Arab Open University at: www.arabou.org
   - Thétys at: www.tethys-univ.org
7. Other MENA-Related Documents
   - MENA Strategy Paper
   - Dialogue across Cultures
   - Urban Development and City Management

- SMExchange Presentation (C. Schmidt and G. Garcia)
- Innovation and Entrepreneurship: The Case of Egypt (S. Delawar)

8. World Bank and MENA-Related Websites
    - MENA Region Website in the World Bank
    - MENA Knowledge Services

9. Knowledge for Development Related Websites
    - Knowledge for Development at: www.worldbank.org/wbi/knowledgefordevelopment/
    - Knowledge Assessment Methodology (KAM) at: www1.worldbank.org/gdln/kam.htm
    - World Bank Institute at: www.worldbank.org/wbi

10. Other Websites:
    - Development Gateway at: www.developmentgateway.org
    - SMExchange at: www.smexchange.org
    - FEMISE at: www.femise.org
    - City of Marseilles at: www.mairie-marseille.fr